2015 IECC®

INTERNATIONAL
Energy Conservation Code®

For Reference
Not to be taken from this room

CODE ALERT!

Sign up now to receive critical code updates and free access to videos, book excerpts and training resources.

Signup is easy, subscribe now! www.iccsafe.org/alerts

2015 International Energy Conservation Code®

First Printing: May 2014
Second Printing: May 2015
Third Printing: January 2016

ISBN: 978-1-60983-486-9 (soft-cover edition)

COPYRIGHT © 2014
by
INTERNATIONAL CODE COUNCIL, INC.

Date of First Publication: May 30, 2014

ALL RIGHTS RESERVED. This 2015 *International Energy Conservation Code®* is a copyrighted work owned by the International Code Council, Inc. Without advance written permission from the copyright owner, no part of this book may be reproduced, distributed or transmitted in any form or by any means, including, without limitation, electronic, optical or mechanical means (by way of example, and not limitation, photocopying, or recording by or in an information storage retrieval system). For information on permission to copy material exceeding fair use, please contact: Publications, 4051 West Flossmoor Road, Country Club Hills, IL 60478. Phone 1-888-ICC-SAFE (422-7233).

Trademarks: "International Code Council," the "International Code Council" logo and the "International Energy Conservation Code" are trademarks of the International Code Council, Inc.

PRINTED IN THE U.S.A.

PREFACE

Introduction

Internationally, code officials recognize the need for a modern, up-to-date energy conservation code addressing the design of energy-efficient building envelopes and installation of energy-efficient mechanical, lighting and power systems through requirements emphasizing performance. The *International Energy Conservation Code®*, in this 2015 edition, is designed to meet these needs through model code regulations that will result in the optimal utilization of fossil fuel and nondepletable resources in all communities, large and small.

This code contains separate provisions for commercial buildings and for low-rise residential buildings (3 stories or less in height above grade). Each set of provisions, IECC—Commercial Provisions and IECC—Residential Provisions, is separately applied to buildings within their respective scopes. Each set of provisions is to be treated separately. Each contains a Scope and Administration chapter, a Definitions chapter, a General Requirements chapter, a chapter containing energy efficiency requirements and existing building provisions applicable to buildings within its scope.

This comprehensive energy conservation code establishes minimum regulations for energy-efficient buildings using prescriptive and performance-related provisions. It is founded on broad-based principles that make possible the use of new materials and new energy-efficient designs. This 2015 edition is fully compatible with all of the *International Codes®* (I-Codes®) published by the International Code Council (ICC)®, including: the *International Building Code®*, *International Existing Building Code®*, *International Fire Code®*, *International Fuel Gas Code®*, *International Green Construction Code®*, *International Mechanical Code®*, *ICC Performance Code®*, *International Plumbing Code®*, *International Private Sewage Disposal Code®*, *International Property Maintenance Code®*, *International Residential Code®*, *International Swimming Pool and Spa Code*™, *International Wildland-Urban Interface Code®* and *International Zoning Code®*.

The *International Energy Conservation Code* provisions provide many benefits, among which is the model code development process that offers an international forum for energy professionals to discuss performance and prescriptive code requirements. This forum provides an excellent arena to debate proposed revisions. This model code also encourages international consistency in the application of provisions.

Development

The first edition of the *International Energy Conservation Code* (1998) was based on the 1995 edition of the *Model Energy Code* promulgated by the Council of American Building Officials (CABO) and included changes approved through the CABO Code Development Procedures through 1997. CABO assigned all rights and responsibilities to the International Code Council and its three statutory members at that time, including Building Officials and Code Administrators International, Inc. (BOCA), International Conference of Building Officials (ICBO) and Southern Building Code Congress International (SBCCI). This 2015 edition presents the code as originally issued, with changes reflected in the 2000 through 2012 editions and with changes approved through the ICC Code Development Process through 2014. A new edition such as this is promulgated every 3 years.

This code is founded on principles intended to establish provisions consistent with the scope of an energy conservation code that adequately conserves energy; provisions that do not unnecessarily increase construction costs; provisions that do not restrict the use of new materials, products or methods of construction; and provisions that do not give preferential treatment to particular types or classes of materials, products or methods of construction.

Adoption

The International Code Council maintains a copyright in all of its codes and standards. Maintaining copyright allows the ICC to fund its mission through sales of books, in both print and electronic formats. The *International Energy Conservation Code* is designed for adoption and use by jurisdictions that recognize and acknowledge the ICC's copyright in the code, and further acknowledge the substantial shared value of the public/private partnership for code development between jurisdictions and the ICC.

The ICC also recognizes the need for jurisdictions to make laws available to the public. All ICC codes and ICC standards, along with the laws of many jurisdictions, are available for free in a nondownloadable form on the ICC's website. Jurisdictions should contact the ICC at adoptions@icc-safe.org to learn how to adopt and distribute laws based on the *International Energy Conservation Code* in a manner that provides necessary access, while maintaining the ICC's copyright.

Maintenance

The *International Energy Conservation Code* is kept up to date through the review of proposed changes submitted by code enforcing officials, industry representatives, design professionals and other interested parties. Proposed changes are carefully considered through an open code development process in which all interested and affected parties may participate.

The contents of this work are subject to change through both the code development cycles and the governmental body that enacts the code into law. For more information regarding the code development process, contact the Codes and Standards Development Department of the International Code Council.

While the development procedure of the *International Energy Conservation Code* assures the highest degree of care, the ICC, its members and those participating in the development of this code do not accept any liability resulting from compliance or noncompliance with the provisions because the ICC does not have the power or authority to police or enforce compliance with the contents of this code. Only the governmental body that enacts the code into law has such authority.

Code Development Committee Responsibilities
(Letter Designations in Front of Section Numbers)

In each code development cycle, proposed changes to the code are considered at the Committee Action Hearings by the applicable International Code Development Committee. The IECC—Commercial Provisions (sections designated with a "C" prior to the section number) are primarily maintained by the Commercial Energy Code Development Committee. The IECC—Residential Provisions (sections designated with an "R" prior to the section number) are maintained by the Residential Energy Code Development Committee. This is designated in the chapter headings by a [CE] and [RE], respectively.

Maintenance responsibilities for the IECC are designated as follows:

[CE] = Commercial Energy Code Development Committee

[RE] = Residential Energy Code Development Committee

For the development of the 2018 edition of the I-Codes, there will be three groups of code development committees and they will meet in separate years. Note that these are tentative groupings.

Group A Codes (Heard in 2015, Code Change Proposals Deadline: January 12, 2015)	Group B Codes (Heard in 2016, Code Change Proposals Deadline: January 11, 2016)	Group C Codes (Heard in 2017, Code Change Proposals Deadline: January 11, 2017)
International Building Code —Fire Safety (Chapters 7, 8, 9, 14, 26) —Means of Egress (Chapters 10, 11, Appendix E) —General (Chapters 2-6, 12, 27-33, Appendices A, B, C, D, K)	Administrative Provisions (Chapter 1 of all codes except IRC and IECC, administrative updates to currently referenced standards, and designated definitions)	International Green Construction Code
International Fuel Gas Code	International Building Code —Structural (Chapters 15-25, Appendices F, G, H, I, J, L, M)	
International Existing Building Code	**International Energy Conservation Code**	
International Mechanical Code	International Fire Code	
International Plumbing Code	International Residential Code —IRC-Building (Chapters 1-10, Appendices E, F, H, J, K, L, M, O, R, S, T, U)	
International Private Sewage Disposal Code	International Wildland-Urban Interface Code	
International Property Maintenance Code		
International Residential Code —IRC-Mechanical (Chapters 12-24) —IRC-Plumbing (Chapters 25-33, Appendices G, I, N, P)		
International Swimming Pool and Spa Code		
International Zoning Code		

Note: Proposed changes to the ICC *Performance Code* will be heard by the code development committee noted in brackets [ ] in the text of the code.

Marginal Markings

Solid vertical lines in the margins within the body of the code indicate a technical change from the requirements of the 2012 edition. Deletion indicators in the form of an arrow (➡) are provided in the margin where an entire section, paragraph, exception or table has been deleted or an item in a list of items or a table has been deleted.

A single asterisk [*] placed in the margin indicates that text or a table has been relocated within the code. A double asterisk [**] placed in the margin indicates that the text or table immediately following it has been relocated there from elsewhere in the code. The following table indicates such relocations in the 2015 edition of the *International Energy Conservation Code*.

2015 LOCATION	2012 LOCATION
C501	C101.4.1
C501.6	C101.4.2
C502 through C504	C101.4.3
C505	C101.4.4
C503.2	C101.4.5
C402.1.1	C101.5.2
C402.3	C402.2.1.1

2015 LOCATION	2012 LOCATION
R501	R101.4.1
R501.6	R101.4.2
R502 through R504	R101.4.3
R505	R101.4.4
R503.2	R101.4.5
R402.1	R101.5.2
R503.1.1.1	R402.3.6

Italicized Terms

Selected terms set forth in Chapter 2, Definitions, are italicized where they appear in code text. Such terms are not italicized where the definition set forth in Chapter 2 does not impart the intended meaning in the use of the term. The terms selected have definitions that the user should read carefully to facilitate better understanding of the code.

EFFECTIVE USE OF THE INTERNATIONAL ENERGY CONSERVATION CODE

The *International Energy Conservation Code* (IECC) is a model code that regulates minimum energy conservation requirements for new buildings. The IECC addresses energy conservation requirements for all aspects of energy uses in both commercial and residential construction, including heating and ventilating, lighting, water heating, and power usage for appliances and building systems.

The IECC is a design document. For example, before one constructs a building, the designer must determine the minimum insulation *R*-values and fenestration *U*-factors for the building exterior envelope. Depending on whether the building is for residential use or for commercial use, the IECC sets forth minimum requirements for exterior envelope insulation, window and door *U*-factors and SHGC ratings, duct insulation, lighting and power efficiency, and water distribution insulation.

Arrangement and Format of the 2015 IECC

The IECC contains two separate sets of provisions—one for commercial buildings and one for residential buildings. Each set of provisions is applied separately to buildings within their scope. The IECC—Commercial Provisions apply to all buildings except for residential buildings three stories or less in height. The IECC—Residential Provisions apply to detached one- and two-family dwellings and multiple single-family dwellings as well as Group R-2, R-3 and R-4 buildings three stories or less in height. These scopes are based on the definitions of "Commercial building" and "Residential building," respectively, in Chapter 2 of each set of provisions. Note that the IECC—Commercial Provisions therefore contain provisions for residential buildings four stories or greater in height. Each set of provisions is divided into five different parts:

Chapters	Subjects
1-2	Administration and definitions
3	Climate zones and general materials requirements
4	Energy efficiency requirements
5	Existing buildings
6	Referenced standards

The following is a chapter-by-chapter synopsis of the scope and intent of the provisions of the *International Energy Conservation Code* and applies to both the commercial and residential energy provisions:

Chapter 1 Scope and Administration. This chapter contains provisions for the application, enforcement and administration of subsequent requirements of the code. In addition to establishing the scope of the code, Chapter 1 identifies which buildings and structures come under its purview. Chapter 1 is largely concerned with maintaining "due process of law" in enforcing the energy conservation criteria contained in the body of this code. Only through careful observation of the administrative provisions can the code official reasonably expect to demonstrate that "equal protection under the law" has been provided.

Chapter 2 Definitions. Chapter 2 is the repository of the definitions of terms used in the body of the code. Codes are technical documents and every word, term and punctuation mark can impact the meaning of the code text and the intended results. The code often uses terms that have a unique meaning in the code and the code meaning can differ substantially from the ordinarily understood meaning of the term as used outside of the code.

The terms defined in Chapter 2 are deemed to be of prime importance in establishing the meaning and intent of the code text. The user of the code should be familiar with and consult this chapter because the definitions are essential to the correct interpretation of the code and the user may not be aware that a term is defined.

Additional definitions regarding climate zones are found in Tables 301.3(1) and (2). These are not listed in Chapter 2.

Where understanding of a term's definition is especially key to or necessary for understanding of a particular code provision, the term is shown in *italics* wherever it appears in the code. This is true only for those terms that have a meaning that is unique to the code. In other words, the generally understood meaning of a term or phrase might not be sufficient or consistent with the meaning prescribed by the code; therefore, it is essential that the code-defined meaning be known.

Guidance regarding tense, gender and plurality of defined terms as well as guidance regarding terms not defined in this code is provided.

Chapter 3 General Requirements. Chapter 3 specifies the climate zones that will serve to establish the exterior design conditions. In addition, Chapter 3 provides interior design conditions that are used as a basis for assumptions in heating and cooling load calculations, and provides basic material requirements for insulation materials and fenestration materials.

Climate has a major impact on the energy use of most buildings. The code establishes many requirements such as wall and roof insulation *R*-values, window and door thermal transmittance requirement (*U*-factors) as well as provisions that affect the mechanical systems based upon the climate where the building is located. This chapter contains information that will be used to properly assign the building location into the correct climate zone and is used as the basis for establishing requirements or elimination of requirements.

Chapter 4 Energy Efficiency. Chapter 4 of each set of provisions contains the technical requirements for energy efficiency.

Commercial Energy Efficiency. Chapter 4 of the IECC—Commercial Provisions contains the energy-efficiency-related requirements for the design and construction of most types of commercial buildings and residential buildings greater than three stories in height above grade. Residential buildings, townhouses and garden apartments three stories or less in height are covered in the IECC—Residential Provisions. This chapter defines requirements for the portions of the building and building systems that impact energy use in new commercial construction and new residential construction greater than three stories in height, and promotes the effective use of energy. The provisions within the chapter promote energy efficiency in the building envelope, the heating and cooling system and the service water heating system of the building.

Residential Energy Efficiency. Chapter 4 of the IECC—Residential Provisions contains the energy-efficiency-related requirements for the design and construction of residential buildings regulated under this code. It should be noted that the definition of a *residential building* in this code is unique for this code. In this code, a *residential building* is a detached one- and two-family dwelling and multiple single-family dwellings as well as R-2, R-3 or R-4 buildings three stories or less in height. All other buildings, including residential buildings greater than three stories in height, are regulated by the energy conservation requirements in the IECC—Commercial Provisions. The applicable portions of a residential building must comply with the provisions within this chapter for energy efficiency. This chapter defines requirements for the portions of the building and building systems that impact energy use in new residential construction and promotes the effective use of energy. The provisions within the chapter promote energy efficiency in the building envelope, the heating and cooling system and the service water heating system of the building.

Chapter 5 Existing Buildings. Chapter 5 of each set of provisions contains the technical energy efficiency requirements for existing buildings. Chapter 5 provisions address the maintenance of buildings in compliance with the code as well as how additions, alterations, repairs and changes of occupancy need to be addressed from the standpoint of energy efficiency. Specific provisions are provided for historic buildings.

Chapter 6 Referenced Standards. The code contains numerous references to standards that are used to regulate materials and methods of construction. Chapter 6 contains a comprehensive list of all standards that are referenced in the code. The standards are part of the code to the extent of the reference to the standard. Compliance with the referenced standard is necessary for compliance with this code. By providing specifically adopted standards, the construction and installation requirements necessary for compliance with the code can be readily determined. The basis for code compliance is, therefore, established and available on an equal basis to the code official, contractor, designer and owner.

Chapter 6 is organized in a manner that makes it easy to locate specific standards. It lists all of the referenced standards, alphabetically, by acronym of the promulgating agency of the standard. Each agency's standards are then listed in either alphabetical or numeric order based upon the standard identification. The list also contains the title of the standard; the edition (date) of the standard referenced; any addenda included as part of the ICC adoption; and the section or sections of this code that reference the standard.

Abbreviations and Notations

The following is a list of common abbreviations and units of measurement used in this code. Some of the abbreviations are for terms defined in Chapter 2. Others are terms used in various tables and text of the code.

AFUE	Annual fuel utilization efficiency
bhp	Brake horsepower (fans)
Btu	British thermal unit
$Btu/h\text{-}ft^2$	Btu per hour per square foot
C-factor	See Chapter 2—Definitions
CDD	Cooling degree days
cfm	Cubic feet per minute
cfm/ft^2	Cubic feet per minute per square foot
ci	Continuous insulation
COP	Coefficient of performance
DCV	Demand control ventilation
°C	Degrees Celsius
°F	Degrees Fahrenheit
DWHR	Drain water heat recovery
DX	Direct expansion
E_c	Combustion efficiency
E_v	Ventilation efficiency
E_t	Thermal efficiency
EER	Energy efficiency ratio
EF	Energy factor
ERI	Energy Rating index
F-factor	See Chapter 2—Definitions

FDD	Fault detection and diagnostics
FEG	Fan efficiency grade
FL	Full load
ft^2	Square foot
gpm	Gallons per minute
HDD	Heating degree days
hp	Horsepower
HSPF	Heating seasonal performance factor
HVAC	Heating, ventilating and air conditioning
IEER	Integrated energy efficiency ratio
IPLV	Integrated Part Load Value
Kg/m^2	Kilograms per square meter
kW	Kilowatt
LPD	Light power density (lighting power allowance)
L/s	Liters per second
Ls	Liner system
m^2	square meters
MERV	Minimum efficiency reporting value
NAECA	National Appliance Energy Conservation Act
NPLV	Nonstandard Part Load Value
Pa	Pascal
PF	Projection factor
pcf	Pounds per cubic foot
psf	Pounds per square foot
PTAC	Packaged terminal air conditioner
PTHP	Packaged terminal heat pump
R-value	See Chapter 2—Definitions
SCOP	Sensible coefficient of performance
SEER	Seasonal energy efficiency ratio
SHGC	Solar Heat Gain Coefficient
SPVAC	Single packaged vertical air conditioner
SPVHP	Single packaged vertical heat pump
SRI	Solar reflectance index
SWHF	Service water heat recovery factor
U-factor	See Chapter 2—Definitions
VAV	Variable air volume
VRF	Variable refrigerant flow
VT	Visible transmittance
W	Watts
w.c.	Water column
w.g.	Water gauge

LEGISLATION

Jurisdictions wishing to adopt the 2015 *International Energy Conservation Code* as an enforceable regulation governing energy-efficient building envelopes and installation of energy-efficient mechanical, lighting and power systems should ensure that certain factual information is included in the adopting legislation at the time adoption is being considered by the appropriate governmental body. The following sample adoption legislation addresses several key elements, including the information required for insertion into the code text.

SAMPLE LEGISLATION FOR ADOPTION OF THE *INTERNATIONAL ENERGY CONSERVATION CODE* ORDINANCE NO._____

A**[N] [ORDINANCE/STATUTE/REGULATION]** of the **[JURISDICTION]** adopting the 2015 edition of the *International Energy Conservation Code*, regulating and governing energy-efficient building envelopes and installation of energy-efficient mechanical, lighting and power systems in the **[JURISDICTION]**; providing for the issuance of permits and collection of fees therefor; repealing **[ORDINANCE/STATUTE/REGULATION]** No. _____ of the **[JURISDICTION]** and all other ordinances or parts of laws in conflict therewith.

The **[GOVERNING BODY]** of the **[JURISDICTION]** does ordain as follows:

Section 1. That a certain document, three (3) copies of which are on file in the office of the **[TITLE OF JURISDICTION'S KEEPER OF RECORDS]** of **[NAME OF JURISDICTION]**, being marked and designated as the *International Energy Conservation Code*, 2015 edition, as published by the International Code Council, be and is hereby adopted as the Energy Conservation Code of the **[JURISDICTION]**, in the State of **[STATE NAME]** for regulating and governing energy-efficient building envelopes and installation of energy-efficient mechanical, lighting and power systems as herein provided; providing for the issuance of permits and collection of fees therefor; and each and all of the regulations, provisions, penalties, conditions and terms of said Energy Conservation Code on file in the office of the **[JURISDICTION]** are hereby referred to, adopted, and made a part hereof, as if fully set out in this legislation, with the additions, insertions, deletions and changes, if any, prescribed in Section 2 of this ordinance.

Section 2. The following sections are hereby revised:

Sections C101.1 and R101.1. Insert: **[NAME OF JURISDICTION]**.

Section 3. That **[ORDINANCE/STATUTE/REGULATION]** No. _____ of **[JURISDICTION]** entitled **[FILL IN HERE THE COMPLETE TITLE OF THE LEGISLATION OR LAWS IN EFFECT AT THE PRESENT TIME SO THAT THEY WILL BE REPEALED BY DEFINITE MENTION]** and all other ordinances or parts of laws in conflict herewith are hereby repealed.

Section 4. That if any section, subsection, sentence, clause or phrase of this legislation is, for any reason, held to be unconstitutional, such decision shall not affect the validity of the remaining portions of this ordinance. The **[GOVERNING BODY]** hereby declares that it would have passed this law, and each section, subsection, clause or phrase thereof, irrespective of the fact that any one or more sections, subsections, sentences, clauses and phrases be declared unconstitutional.

Section 5. That nothing in this legislation or in the Energy Conservation Code hereby adopted shall be construed to affect any suit or proceeding impending in any court, or any rights acquired, or liability incurred, or any cause or causes of action acquired or existing, under any act or ordinance hereby repealed as cited in Section 3 of this law; nor shall any just or legal right or remedy of any character be lost, impaired or affected by this legislation.

Section 6. That the **[JURISDICTION'S KEEPER OF RECORDS]** is hereby ordered and directed to cause this legislation to be published. (An additional provision may be required to direct the number of times the legislation is to be published and to specify that it is to be in a newspaper in general circulation. Posting may also be required.)

Section 7. That this law and the rules, regulations, provisions, requirements, orders and matters established and adopted hereby shall take effect and be in full force and effect **[TIME PERIOD]** from and after the date of its final passage and adoption.

TABLE OF CONTENTS

IECC—COMMERCIAL PROVISIONS.....C-1

CHAPTER 1 SCOPE AND ADMINISTRATION C-3

CHAPTER 2 DEFINITIONS C-7

CHAPTER 3 GENERAL REQUIREMENTS C-13

CHAPTER 4 COMMERCIAL ENERGY EFFICIENCY C-31

CHAPTER 5 EXISTING BUILDINGS C-95

CHAPTER 6 REFERENCED STANDARDS C-99

INDEX C-107

IECC—RESIDENTIAL PROVISIONS..... R-1

CHAPTER 1 SCOPE AND ADMINISTRATION R-3

CHAPTER 2 DEFINITIONS R-7

CHAPTER 3 GENERAL REQUIREMENTS R-11

CHAPTER 4 RESIDENTIAL ENERGY EFFICIENCY R-29

CHAPTER 5 EXISTING BUILDINGS R-43

CHAPTER 6 REFERENCED STANDARDS..... R-45

APPENDIX RA RECOMMENDED PROCEDURE FOR WORST-CASE TESTING OF ATMOSPHERIC VENTING SYSTEMS UNDER R402.4 OR R405 CONDITIONS $\leq 5ACH_{50}$ R-49

APPENDIX RB SOLAR-READY PROVISIONS—DETACHED ONE- AND TWO-FAMILY DWELLINGS, MULTIPLE SINGLE-FAMILY DWELLINGS (TOWNHOUSES) R-51

INDEX R-53

IECC—COMMERCIAL PROVISIONS

TABLE OF CONTENTS

CHAPTER 1 SCOPE AND ADMINISTRATION C-3

PART 1—SCOPE AND APPLICATION........... C-3
Section
C101 Scope and General Requirements............ C-3
C102 Alternate Materials—Method of Construction, Design or Insulating Systems.............................. C-3

PART 2—ADMINISTRATION AND ENFORCEMENT C-3
C103 Construction Documents C-3
C104 Inspections C-4
C105 Validity................................. C-5
C106 Referenced Standards..................... C-5
C107 Fees.................................... C-5
C108 Stop Work Order C-5
C109 Board of Appeals C-5

CHAPTER 2 DEFINITIONS C-7
Section
C201 General................................ C-7
C202 General Definitions C-7

CHAPTER 3 GENERAL REQUIREMENTS..... C-13
Section
C301 Climate Zones........................ C-13
C302 Design Conditions C-28
C303 Materials, Systems and Equipment C-28

CHAPTER 4 COMMERCIAL ENERGY EFFICIENCY C-31
Section
C401 General............................... C-31
C402 Building Envelope Requirements C-31
C403 Building Mechanical Systems C-40
C404 Service Water Heating................... C-68
C405 Electrical Power and Lighting Systems...... C-71
C406 Additional Efficiency Package Options...... C-83
C407 Total Building Performance............... C-84
C408 System Commissioning C-90

CHAPTER 5 EXISTING BUILDINGS C-95
Section
C501 General............................... C-95
C502 Additions C-95
C503 Alterations C-96
C504 Repairs C-96
C505 Change of Occupancy or Use.............. C-97

CHAPTER 6 REFERENCED STANDARDS C-99

INDEX...................................... C-107

CHAPTER 1 [CE]

SCOPE AND ADMINISTRATION

PART 1—SCOPE AND APPLICATION

SECTION C101
SCOPE AND GENERAL REQUIREMENTS

C101.1 Title. This code shall be known as the *International Energy Conservation Code* of **[NAME OF JURISDICTION]**, and shall be cited as such. It is referred to herein as "this code."

C101.2 Scope. This code applies to *commercial buildings* and the buildings' sites and associated systems and equipment.

C101.3 Intent. This code shall regulate the design and construction of buildings for the use and conservation of energy over the life of each building. This code is intended to provide flexibility to permit the use of innovative approaches and techniques to achieve this objective. This code is not intended to abridge safety, health or environmental requirements contained in other applicable codes or ordinances.

C101.4 Applicability. Where, in any specific case, different sections of this code specify different materials, methods of construction or other requirements, the most restrictive shall govern. Where there is a conflict between a general requirement and a specific requirement, the specific requirement shall govern.

C101.4.1 Mixed occupancy. Where a building includes both *residential* and *commercial* occupancies, each occupancy shall be separately considered and meet the applicable provisions of IECC—Commercial Provisions or IECC—Residential Provisions.

C101.5 Compliance. *Residential buildings* shall meet the provisions of IECC—Residential Provisions. *Commercial buildings* shall meet the provisions of IECC—Commercial Provisions.

C101.5.1 Compliance materials. The *code official* shall be permitted to approve specific computer software, worksheets, compliance manuals and other similar materials that meet the intent of this code.

SECTION C102
ALTERNATE MATERIALS—METHOD OF CONSTRUCTION, DESIGN OR INSULATING SYSTEMS

C102.1 General. This code is not intended to prevent the use of any material, method of construction, design or insulating system not specifically prescribed herein, provided that such construction, design or insulating system has been *approved* by the *code official* as meeting the intent of this code.

C102.1.1 Above code programs. The *code official* or other authority having jurisdiction shall be permitted to deem a national, state or local energy efficiency program to exceed the energy efficiency required by this code. Buildings *approved* in writing by such an energy efficiency program shall be considered in compliance with this code. The requirements identified as "mandatory" in Chapter 4 shall be met.

PART 2—ADMINISTRATION AND ENFORCEMENT

SECTION C103
CONSTRUCTION DOCUMENTS

C103.1 General. Construction documents and other supporting data shall be submitted in one or more sets with each application for a permit. The construction documents shall be prepared by a registered design professional where required by the statutes of the jurisdiction in which the project is to be constructed. Where special conditions exist, the *code official* is authorized to require necessary construction documents to be prepared by a registered design professional.

> **Exception:** The *code official* is authorized to waive the requirements for construction documents or other supporting data if the *code official* determines they are not necessary to confirm compliance with this code.

C103.2 Information on construction documents. Construction documents shall be drawn to scale upon suitable material. Electronic media documents are permitted to be submitted where *approved* by the *code official*. Construction documents shall be of sufficient clarity to indicate the location, nature and extent of the work proposed, and show in sufficient detail pertinent data and features of the building, systems and equipment as herein governed. Details shall include, but are not limited to, the following as applicable:

1. Insulation materials and their *R*-values.
2. Fenestration *U*-factors and solar heat gain coefficients (SHGCs).
3. Area-weighted *U*-factor and solar heat gain coefficient (SHGC) calculations.
4. Mechanical system design criteria.
5. Mechanical and service water heating system and equipment types, sizes and efficiencies.
6. Economizer description.
7. Equipment and system controls.
8. Fan motor horsepower (hp) and controls.
9. Duct sealing, duct and pipe insulation and location.
10. Lighting fixture schedule with wattage and control narrative.
11. Location of *daylight* zones on floor plans.
12. Air sealing details.

C103.2.1 Building thermal envelope depiction. The *building's thermal envelope* shall be represented on the construction drawings.

C103.3 Examination of documents. The *code official* shall examine or cause to be examined the accompanying construction documents and shall ascertain whether the construction indicated and described is in accordance with the requirements of this code and other pertinent laws or ordinances. The *code official* is authorized to utilize a registered design professional, or other *approved* entity not affiliated with the building design or construction, in conducting the review of the plans and specifications for compliance with the code.

C103.3.1 Approval of construction documents. When the *code official* issues a permit where construction documents are required, the construction documents shall be endorsed in writing and stamped "Reviewed for Code Compliance." Such *approved* construction documents shall not be changed, modified or altered without authorization from the *code official*. Work shall be done in accordance with the *approved* construction documents.

One set of construction documents so reviewed shall be retained by the *code official*. The other set shall be returned to the applicant, kept at the site of work and shall be open to inspection by the *code official* or a duly authorized representative.

C103.3.2 Previous approvals. This code shall not require changes in the construction documents, construction or designated occupancy of a structure for which a lawful permit has been heretofore issued or otherwise lawfully authorized, and the construction of which has been pursued in good faith within 180 days after the effective date of this code and has not been abandoned.

C103.3.3 Phased approval. The *code official* shall have the authority to issue a permit for the construction of part of an energy conservation system before the construction documents for the entire system have been submitted or *approved*, provided adequate information and detailed statements have been filed complying with all pertinent requirements of this code. The holders of such permit shall proceed at their own risk without assurance that the permit for the entire energy conservation system will be granted.

C103.4 Amended construction documents. Changes made during construction that are not in compliance with the *approved* construction documents shall be resubmitted for approval as an amended set of construction documents.

C103.5 Retention of construction documents. One set of *approved* construction documents shall be retained by the *code official* for a period of not less than 180 days from date of completion of the permitted work, or as required by state or local laws.

SECTION C104
INSPECTIONS

C104.1 General. Construction or work for which a permit is required shall be subject to inspection by the *code official* or his or her designated agent, and such construction or work shall remain accessible and exposed for inspection purposes until *approved*. It shall be the duty of the permit applicant to cause the work to remain accessible and exposed for inspection purposes. Neither the *code official* nor the jurisdiction shall be liable for expense entailed in the removal or replacement of any material, product, system or building component required to allow inspection to validate compliance with this code.

C104.2 Required inspections. The *code official* or his or her designated agent, upon notification, shall make the inspections set forth in Sections C104.2.1 through C104.2.6.

C104.2.1 Footing and foundation inspection. Inspections associated with footings and foundations shall verify compliance with the code as to *R*-value, location, thickness, depth of burial and protection of insulation as required by the code and *approved* plans and specifications.

C104.2.2 Framing and rough-in inspection. Inspections at framing and rough-in shall be made before application of interior finish and shall verify compliance with the code as to types of insulation and corresponding *R*-values and their correct location and proper installation; fenestration properties (*U*-factor, SHGC and VT) and proper installation; and air leakage controls as required by the code and approved plans and specifications.

C104.2.3 Plumbing rough-in inspection. Inspections at plumbing rough-in shall verify compliance as required by the code and *approved* plans and specifications as to types of insulation and corresponding *R*-values and protection; required controls; and required heat traps.

C104.2.4 Mechanical rough-in inspection. Inspections at mechanical rough-in shall verify compliance as required by the code and *approved* plans and specifications as to installed HVAC equipment type and size; required controls, system insulation and corresponding *R*-value; system and damper air leakage; and required energy recovery and economizers.

C104.2.5 Electrical rough-in inspection. Inspections at electrical rough-in shall verify compliance as required by the code and *approved* plans and specifications as to installed lighting systems, components and controls; and installation of an electric meter for each dwelling unit.

C104.2.6 Final inspection. The building shall have a final inspection and shall not be occupied until *approved*. The final inspection shall include verification of the installation and proper operation of all required building controls, and documentation verifying activities associated with required *building commissioning* have been conducted and findings of noncompliance corrected. Buildings, or portions thereof, shall not be considered for a final inspection until the *code official* has received a letter of transmittal from the building owner acknowledging that the building owner has received the Preliminary Commissioning Report as required in Section C408.2.4.

C104.3 Reinspection. A building shall be reinspected when determined necessary by the *code official*.

C104.4 Approved inspection agencies. The *code official* is authorized to accept reports of third-party inspection agencies not affiliated with the building design or construction, provided such agencies are *approved* as to qualifications and reliability relevant to the building components and systems they are inspecting.

C104.5 Inspection requests. It shall be the duty of the holder of the permit or their duly authorized agent to notify the *code official* when work is ready for inspection. It shall be the duty of the permit holder to provide access to and means for inspections of such work that are required by this code.

C104.6 Reinspection and testing. Where any work or installation does not pass an initial test or inspection, the necessary corrections shall be made to achieve compliance with this code. The work or installation shall then be resubmitted to the *code official* for inspection and testing.

C104.7 Approval. After the prescribed tests and inspections indicate that the work complies in all respects with this code, a notice of approval shall be issued by the *code official*.

C104.7.1 Revocation. The *code official* is authorized to, in writing, suspend or revoke a notice of approval issued under the provisions of this code wherever the certificate is issued in error, or on the basis of incorrect information supplied, or where it is determined that the *building* or structure, premise, or portion thereof is in violation of any ordinance or regulation or any of the provisions of this code.

SECTION C105
VALIDITY

C105.1 General. If a portion of this code is held to be illegal or void, such a decision shall not affect the validity of the remainder of this code.

SECTION C106
REFERENCED STANDARDS

C106.1 Referenced codes and standards. The codes and standards referenced in this code shall be those listed in Chapter 6, and such codes and standards shall be considered as part of the requirements of this code to the prescribed extent of each such reference and as further regulated in Sections C106.1.1 and C106.1.2.

C106.1.1 Conflicts. Where conflicts occur between provisions of this code and referenced codes and standards, the provisions of this code shall apply.

C106.1.2 Provisions in referenced codes and standards. Where the extent of the reference to a referenced code or standard includes subject matter that is within the scope of this code, the provisions of this code, as applicable, shall take precedence over the provisions in the referenced code or standard.

C106.2 Application of references. References to chapter or section numbers, or to provisions not specifically identified by number, shall be construed to refer to such chapter, section or provision of this code.

C106.3 Other laws. The provisions of this code shall not be deemed to nullify any provisions of local, state or federal law.

SECTION C107
FEES

C107.1 Fees. A permit shall not be issued until the fees prescribed in Section C107.2 have been paid, nor shall an amendment to a permit be released until the additional fee, if any, has been paid.

C107.2 Schedule of permit fees. A fee for each permit shall be paid as required, in accordance with the schedule as established by the applicable governing authority.

C107.3 Work commencing before permit issuance. Any person who commences any work before obtaining the necessary permits shall be subject to an additional fee established by the *code official* that shall be in addition to the required permit fees.

C107.4 Related fees. The payment of the fee for the construction, *alteration*, removal or demolition of work done in connection to or concurrently with the work or activity authorized by a permit shall not relieve the applicant or holder of the permit from the payment of other fees that are prescribed by law.

C107.5 Refunds. The *code official* is authorized to establish a refund policy.

SECTION C108
STOP WORK ORDER

C108.1 Authority. Where the *code official* finds any work regulated by this code being performed in a manner either contrary to the provisions of this code or dangerous or unsafe, the *code official* is authorized to issue a stop work order.

C108.2 Issuance. The stop work order shall be in writing and shall be given to the owner of the property involved, the owner's authorized agent, or to the person doing the work. Upon issuance of a stop work order, the cited work shall immediately cease. The stop work order shall state the reason for the order and the conditions under which the cited work will be permitted to resume.

C108.3 Emergencies. Where an emergency exists, the *code official* shall not be required to give a written notice prior to stopping the work.

C108.4 Failure to comply. Any person who shall continue any work after having been served with a stop work order, except such work as that person is directed to perform to remove a violation or unsafe condition, shall be liable to a fine as set by the applicable governing authority.

SECTION C109
BOARD OF APPEALS

C109.1 General. In order to hear and decide appeals of orders, decisions or determinations made by the *code official* relative to the application and interpretation of this code, there shall be and is hereby created a board of appeals. The *code official* shall be an ex officio member of said board but shall not have a vote on any matter before the board. The board of appeals shall be appointed by the governing body and shall hold office at its pleasure. The board shall adopt rules of procedure for conducting its business, and shall render all decisions and findings in writing to the appellant with a duplicate copy to the *code official*.

C109.2 Limitations on authority. An application for appeal shall be based on a claim that the true intent of this code or the rules legally adopted thereunder have been incorrectly interpreted, the provisions of this code do not fully apply or an equally good or better form of construction is proposed. The board shall not have authority to waive requirements of this code.

C109.3 Qualifications. The board of appeals shall consist of members who are qualified by experience and training and are not employees of the jurisdiction.

CHAPTER 2 [CE]
DEFINITIONS

SECTION C201
GENERAL

C201.1 Scope. Unless stated otherwise, the following words and terms in this code shall have the meanings indicated in this chapter.

C201.2 Interchangeability. Words used in the present tense include the future; words in the masculine gender include the feminine and neuter; the singular number includes the plural and the plural includes the singular.

C201.3 Terms defined in other codes. Terms that are not defined in this code but are defined in the *International Building Code*, *International Fire Code*, *International Fuel Gas Code*, *International Mechanical Code*, *International Plumbing Code* or the *International Residential Code* shall have the meanings ascribed to them in those codes.

C201.4 Terms not defined. Terms not defined by this chapter shall have ordinarily accepted meanings such as the context implies.

SECTION C202
GENERAL DEFINITIONS

ABOVE-GRADE WALL. See "Wall, above-grade."

ACCESSIBLE. Admitting close approach as a result of not being guarded by locked doors, elevation or other effective means (see "Readily *accessible*").

ADDITION. An extension or increase in the *conditioned space* floor area or height of a building or structure.

AIR BARRIER. Materials assembled and joined together to provide a barrier to air leakage through the building envelope. An air barrier may be a single material or a combination of materials.

AIR CURTAIN. A device, installed at the building entrance, that generates and discharges a laminar air stream intended to prevent the infiltration of external, unconditioned air into the conditioned spaces, or the loss of interior, conditioned air to the outside.

ALTERATION. Any construction, retrofit or renovation to an existing structure other than repair or addition that requires a permit. Also, a change in a building, electrical, gas, mechanical or plumbing system that involves an extension, addition or change to the arrangement, type or purpose of the original installation that requires a permit.

APPROVED. Approval by the *code official* as a result of investigation and tests conducted by him or her, or by reason of accepted principles or tests by nationally recognized organizations.

APPROVED AGENCY. An established and recognized agency regularly engaged in conducting tests or furnishing inspection services, when such agency has been approved by the *code official*.

AUTOMATIC. Self-acting, operating by its own mechanism when actuated by some impersonal influence, as, for example, a change in current strength, pressure, temperature or mechanical configuration (see "Manual").

BELOW-GRADE WALL. See "Wall, below-grade."

BOILER, MODULATING. A boiler that is capable of more than a single firing rate in response to a varying temperature or heating load.

BOILER SYSTEM. One or more boilers, their piping and controls that work together to supply steam or hot water to heat output devices remote from the boiler.

BUBBLE POINT. The refrigerant liquid saturation temperature at a specified pressure.

BUILDING. Any structure used or intended for supporting or sheltering any use or occupancy, including any mechanical systems, service water heating systems and electric power and lighting systems located on the building site and supporting the building.

BUILDING COMMISSIONING. A process that verifies and documents that the selected building systems have been designed, installed, and function according to the owner's project requirements and construction documents, and to minimum code requirements.

BUILDING ENTRANCE. Any door, set of doors, doorway, or other form of portal that is used to gain access to the building from the outside by the public.

BUILDING SITE. A contiguous area of land that is under the ownership or control of one entity.

BUILDING THERMAL ENVELOPE. The basement walls, exterior walls, floor, roof and any other building elements that enclose *conditioned space* or provide a boundary between *conditioned space* and exempt or unconditioned space.

***C*-FACTOR (THERMAL CONDUCTANCE).** The coefficient of heat transmission (surface to surface) through a building component or assembly, equal to the time rate of heat flow per unit area and the unit temperature difference between the warm side and cold side surfaces (Btu/h · ft^2 · °F) [W/(m^2 · K)].

CIRCULATING HOT WATER SYSTEM. A specifically designed water distribution system where one or more pumps are operated in the service hot water piping to circulate heated water from the water-heating equipment to the fixture supply and back to the water-heating equipment.

CLIMATE ZONE. A geographical region based on climatic criteria as specified in this code.

DEFINITIONS

CODE OFFICIAL. The officer or other designated authority charged with the administration and enforcement of this code, or a duly authorized representative.

COEFFECINT OF PERFORMANCE (COP) – COOLING. The ratio of the rate of heat input, in consistent units, for a complete refrigerating system or some specific portion of that system under designated operating conditions.

COEFFICIENT OF PERFORMANCE (COP) – HEATING. The ratio of the rate of heat delivered to the rate of energy input, in consistent units, for a complete heat pump system, including the compressor and, if applicable, auxiliary heat, under designated operating conditions.

COMMERCIAL BUILDING. For this code, all buildings that are not included in the definition of "Residential building."

COMPUTER ROOM. A room whose primary function is to house equipment for the processing and storage of electronic data and that has a design electronic data equipment power density exceeding 20 watts per square foot of conditioned floor area.

CONDENSING UNIT. A factory-made assembly of refrigeration components designed to compress and liquefy a specific refrigerant. The unit consists of one or more refrigerant compressors, refrigerant condensers (air-cooled, evaporatively cooled, or water-cooled), condenser fans and motors (where used) and factory-supplied accessories.

CONDITIONED FLOOR AREA. The horizontal projection of the floors associated with the *conditioned space*.

CONDITIONED SPACE. An area, room or space that is enclosed within the building thermal envelope and is directly or indirectly heated or cooled. Spaces are indirectly heated or cooled where they communicate through openings with conditioned spaces, where they are separated from conditioned spaces by uninsulated walls, floors or ceilings, or where they contain uninsulated ducts, piping or other sources of heating or cooling.

CONTINUOUS AIR BARRIER. A combination of materials and assemblies that restrict or prevent the passage of air through the building thermal envelope.

CONTINUOUS INSULATION (ci). Insulating material that is continuous across all structural members without thermal bridges other than fasteners and service openings. It is installed on the interior or exterior or is integral to any opaque surface of the building envelope.

CRAWL SPACE WALL. The opaque portion of a wall that encloses a crawl space and is partially or totally below grade.

CURTAIN WALL. Fenestration products used to create an external nonload-bearing wall that is designed to separate the exterior and interior environments.

DAYLIGHT RESPONSIVE CONTROL. A device or system that provides automatic control of electric light levels based on the amount of daylight in a space.

DAYLIGHT ZONE. That portion of a building's interior floor area that is illuminated by natural light.

DEMAND CONTROL VENTILATION (DCV). A ventilation system capability that provides for the automatic reduction of outdoor air intake below design rates when the actual occupancy of spaces served by the system is less than design occupancy.

DEMAND RECIRCULATION WATER SYSTEM. A water distribution system where pumps prime the service hot water piping with heated water upon demand for hot water.

DUCT. A tube or conduit utilized for conveying air. The air passages of self-contained systems are not to be construed as air ducts.

DUCT SYSTEM. A continuous passageway for the transmission of air that, in addition to ducts, includes duct fittings, dampers, plenums, fans and accessory air-handling equipment and appliances.

[B] DWELLING UNIT. A single unit providing complete independent living facilities for one or more persons, including permanent provisions for living, sleeping, eating, cooking and sanitation.

DYNAMIC GLAZING. Any fenestration product that has the fully reversible ability to change its performance properties, including *U*-factor, solar heat gain coefficient (SHGC), or visible transmittance (VT).

ECONOMIZER, AIR. A duct and damper arrangement and automatic control system that allows a cooling system to supply outside air to reduce or eliminate the need for mechanical cooling during mild or cold weather.

ECONOMIZER, WATER. A system where the supply air of a cooling system is cooled indirectly with water that is itself cooled by heat or mass transfer to the environment without the use of mechanical cooling.

ENCLOSED SPACE. A volume surrounded by solid surfaces such as walls, floors, roofs, and openable devices such as doors and operable windows.

ENERGY ANALYSIS. A method for estimating the annual energy use of the *proposed design* and *standard reference design* based on estimates of energy use.

ENERGY COST. The total estimated annual cost for purchased energy for the building functions regulated by this code, including applicable demand charges.

[M] ENERGY RECOVERY VENTILATION SYSTEM. Systems that employ air-to-air heat exchangers to recover energy from exhaust air for the purpose of preheating, precooling, humidifying or dehumidifying outdoor ventilation air prior to supplying the air to a space, either directly or as part of an HVAC system.

ENERGY SIMULATION TOOL. An *approved* software program or calculation-based methodology that projects the annual energy use of a building.

ENTRANCE DOOR. Fenestration products used for ingress, egress and access in nonresidential buildings, including, but not limited to, exterior entrances that utilize latching hardware and automatic closers and contain over 50-percent glass specifically designed to withstand heavy use and possibly abuse.

EQUIPMENT ROOM. A space that contains either electrical equipment, mechanical equipment, machinery, water pumps or hydraulic pumps that are a function of the building's services.

EXTERIOR WALL. Walls including both above-grade walls and basement walls.

FAN BRAKE HORSEPOWER (BHP). The horsepower delivered to the fan's shaft. Brake horsepower does not include the mechanical drive losses (belts, gears, etc.).

FAN EFFICIENCY GRADE (FEG). A numerical rating identifying the fan's aerodynamic ability to convert shaft power, or impeller power in the case of a direct-driven fan, to air power.

FAN SYSTEM BHP. The sum of the fan brake horsepower of all fans that are required to operate at fan system design conditions to supply air from the heating or cooling source to the *conditioned spaces* and return it to the source or exhaust it to the outdoors.

FAN SYSTEM DESIGN CONDITIONS. Operating conditions that can be expected to occur during normal system operation that result in the highest supply fan airflow rate to conditioned spaces served by the system.

FAN SYSTEM MOTOR NAMEPLATE HP. The sum of the motor nameplate horsepower of all fans that are required to operate at design conditions to supply air from the heating or cooling source to the *conditioned spaces* and return it to the source or exhaust it to the outdoors.

FENESTRATION. Products classified as either vertical fenestration or skylights.

   Skylight. Glass or other transparent or translucent glazing material installed at a slope of less than 60 degrees (1.05 rad) from horizontal.

   Vertical fenestration. Windows (fixed or moveable), opaque doors, glazed doors, glazed block and combination opaque/glazed doors composed of glass or other transparent or translucent glazing materials and installed at a slope of at least 60 degrees (1.05 rad) from horizontal.

FENESTRATION PRODUCT, FIELD-FABRICATED. A fenestration product whose frame is made at the construction site of standard dimensional lumber or other materials that were not previously cut, or otherwise formed with the specific intention of being used to fabricate a fenestration product or exterior door. Field fabricated does not include site-built fenestration.

FENESTRATION PRODUCT, SITE-BUILT. A fenestration designed to be made up of field-glazed or field-assembled units using specific factory cut or otherwise factory-formed framing and glazing units. Examples of site-built fenestration include storefront systems, curtain walls, and atrium roof systems.

F-FACTOR. The perimeter heat loss factor for slab-on-grade floors (Btu/h · ft · °F) [W/(m · K)].

FLOOR AREA, NET. The actual occupied area not including unoccupied accessory areas such as corridors, stairways, toilet rooms, mechanical rooms and closets.

GENERAL LIGHTING. Lighting that provides a substantially uniform level of illumination throughout an area. General lighting shall not include decorative lighting or lighting that provides a dissimilar level of illumination to serve a specialized application or feature within such area.

GENERAL PURPOSE ELECTRIC MOTOR (SUBTYPE I). A motor that is designed in standard ratings with either of the following:

   1. Standard operating characteristics and standard mechanical construction for use under usual service conditions, such as those specified in NEMA MG1, paragraph 14.02, "Usual Service Conditions," and without restriction to a particular application or type of application.

   2. Standard operating characteristics or standard mechanical construction for use under unusual service conditions, such as those specified in NEMA MG1, paragraph 14.03, "Unusual Service Conditions," or for a particular type of application, and that can be used in most general purpose applications.

General purpose electric motors (Subtype I) are constructed in NEMA T-frame sizes or IEC metric equivalent, starting at 143T.

GENERAL PURPOSE ELECTRIC MOTOR (SUBTYPE II). A motor incorporating the design elements of a general purpose electric motor (Subtype I) that is configured as one of the following:

   1. A U-frame motor.

   2. A Design C motor.

   3. A close-coupled pump motor.

   4. A footless motor.

   5. A vertical, solid-shaft, normal-thrust motor (as tested in a horizontal configuration).

   6. An 8-pole motor (900 rpm).

   7. A polyphase motor with voltage of not more than 600 volts (other than 230 or 460 volts).

GREENHOUSE. A structure or a thermally isolated area of a building that maintains a specialized sunlit environment exclusively used for, and essential to, the cultivation, protection or maintenance of plants.

HEAT TRAP. An arrangement of piping and fittings, such as elbows, or a commercially available heat trap that prevents thermosyphoning of hot water during standby periods.

HEATED SLAB. Slab-on-grade construction in which the heating elements, hydronic tubing, or hot air distribution system is in contact with, or placed within or under, the slab.

HIGH SPEED DOOR. A nonswinging door used primarily to facilitate vehicular access or material transportation, with a minimum opening rate of 32 inches (813 mm) per second, a minimum closing rate of 24 inches (610 mm) per second and that includes an automatic-closing device.

DEFINITIONS

HISTORIC BUILDING. Any building or structure that is one or more of the following:

1. Listed, or certified as eligible for listing by the State Historic Preservation Officer or the Keeper of the National Register of Historic Places, in the National Register of Historic Places.

2. Designated as historic under an applicable state or local law.

3. Certified as a contributing resource within a National Register-listed, state-designated or locally designated historic district.

HUMIDISTAT. A regulatory device, actuated by changes in humidity, used for automatic control of relative humidity.

INFILTRATION. The uncontrolled inward air leakage into a building caused by the pressure effects of wind or the effect of differences in the indoor and outdoor air density or both.

INTEGRATED PART LOAD VALUE (IPLV). A single-number figure of merit based on part-load EER, COP or kW/ton expressing part-load efficiency for air-conditioning and heat pump equipment on the basis of weighted operation at various load capacities for equipment.

LABELED. Equipment, materials or products to which have been affixed a label, seal, symbol or other identifying mark of a nationally recognized testing laboratory, inspection agency or other organization concerned with product evaluation that maintains periodic inspection of the production of the above-labeled items and whose labeling indicates either that the equipment, material or product meets identified standards or has been tested and found suitable for a specified purpose.

LINER SYSTEM (Ls). A system that includes the following:

1. A continuous vapor barrier liner membrane that is installed below the purlins and that is uninterrupted by framing members.

2. An uncompressed, unfaced insulation resting on top of the liner membrane and located between the purlins.

For multilayer installations, the last rated *R-value* of insulation is for unfaced insulation draped over purlins and then compressed when the metal roof panels are attached.

LISTED. Equipment, materials, products or services included in a list published by an organization acceptable to the *code official* and concerned with evaluation of products or services that maintains periodic inspection of production of *listed* equipment or materials or periodic evaluation of services and whose listing states either that the equipment, material, product or service meets identified standards or has been tested and found suitable for a specified purpose.

LOW-SLOPED ROOF. A roof having a slope less than 2 units vertical in 12 units horizontal.

LOW-VOLTAGE DRY-TYPE DISTRIBUTION TRANSFORMER. A transformer that is air-cooled, does not use oil as a coolant, has an input voltage less than or equal to 600 volts and is rated for operation at a frequency of 60 hertz.

LOW-VOLTAGE LIGHTING. Lighting equipment powered through a transformer such as a cable conductor, a rail conductor and track lighting.

MANUAL. Capable of being operated by personal intervention (see "Automatic").

NAMEPLATE HORSEPOWER. The nominal motor horsepower rating stamped on the motor nameplate.

NONSTANDARD PART LOAD VALUE (NPLV). A single-number part-load efficiency figure of merit calculated and referenced to conditions other than IPLV conditions, for units that are not designed to operate at AHRI standard rating conditions.

OCCUPANT SENSOR CONTROL. An automatic control device or system that detects the presence or absence of people within an area and causes lighting, equipment or appliances to be regulated accordingly.

ON-SITE RENEWABLE ENERGY. Energy derived from solar radiation, wind, waves, tides, landfill gas, biomass or the internal heat of the earth. The energy system providing on-site renewable energy shall be located on the project site.

OPAQUE DOOR. A door that is not less than 50-percent opaque in surface area.

POWERED ROOF/WALL VENTILATORS. A fan consisting of a centrifugal or axial impeller with an integral driver in a weather-resistant housing and with a base designed to fit, usually by means of a curb, over a wall or roof opening.

PROPOSED DESIGN. A description of the proposed building used to estimate annual energy use for determining compliance based on total building performance.

RADIANT HEATING SYSTEM. A heating system that transfers heat to objects and surfaces within a conditioned space, primarily by infrared radiation.

READILY ACCESSIBLE. Capable of being reached quickly for operation, renewal or inspection without requiring those to whom ready access is requisite to climb over or remove obstacles or to resort to portable ladders or access equipment (see "*Accessible*").

REFRIGERANT DEW POINT. The refrigerant vapor saturation temperature at a specified pressure.

REFRIGERATED WAREHOUSE COOLER. An enclosed storage space capable of being refrigerated to temperatures above 32°F (0°C), that can be walked into and has a total chilled storage area of not less than 3,000 square feet (279 m^2).

REFRIGERATED WAREHOUSE FREEZER. An enclosed storage space capable of being refrigerated to temperatures at or below 32°F (0°C), that can be walked into and has a total chilled storage area of not less than 3,000 square feet (279 m^2).

REFRIGERATION SYSTEM, LOW TEMPERATURE. Systems for maintaining food product in a frozen state in refrigeration applications.

REFRIGERATION SYSTEM, MEDIUM TEMPERATURE. Systems for maintaining food product above freezing in refrigeration applications.

DEFINITIONS

REGISTERED DESIGN PROFESSIONAL. An individual who is registered or licensed to practice their respective design profession as defined by the statutory requirements of the professional registration laws of the state or jurisdiction in which the project is to be constructed.

REPAIR. The reconstruction or renewal of any part of an existing building for the purpose of its maintenance or to correct damage.

REROOFING. The process of recovering or replacing an existing roof covering. See "Roof recover" and "Roof replacement."

RESIDENTIAL BUILDING. For this code, includes detached one- and two-family dwellings and multiple single-family dwellings (townhouses) as well as Group R-2, R-3 and R-4 buildings three stories or less in height above grade plane.

ROOF ASSEMBLY. A system designed to provide weather protection and resistance to design loads. The system consists of a roof covering and roof deck or a single component serving as both the roof covering and the roof deck. A roof assembly includes the roof covering, underlayment, roof deck, insulation, vapor retarder and interior finish.

ROOF RECOVER. The process of installing an additional roof covering over an existing roof covering without removing the existing roof covering.

ROOF REPAIR. Reconstruction or renewal of any part of an existing roof for the purpose of its maintenance.

ROOF REPLACMENT. The process of removing the existing roof covering, repairing any damaged substrate and installing a new roof covering.

ROOFTOP MONITOR. A raised section of a roof containing vertical fenestration along one or more sides.

R-VALUE (THERMAL RESISTANCE). The inverse of the time rate of heat flow through a body from one of its bounding surfaces to the other surface for a unit temperature difference between the two surfaces, under steady state conditions, per unit area ($h \cdot ft^2 \cdot °F/Btu$) [($m^2 \cdot K$)/W].

SATURATED CONDENSING TEMPERATURE. The saturation temperature corresponding to the measured refrigerant pressure at the condenser inlet for single component and azeotropic refrigerants, and the arithmetic average of the dew point and *bubble point* temperatures corresponding to the refrigerant pressure at the condenser entrance for zeotropic refrigerants.

SCREW LAMP HOLDERS. A lamp base that requires a screw-in-type lamp, such as a compact-fluorescent, incandescent or tungsten-halogen bulb.

SERVICE WATER HEATING. Supply of hot water for purposes other than comfort heating.

[B] SLEEPING UNIT. A room or space in which people sleep, which can also include permanent provisions for living, eating, and either sanitation or kitchen facilities but not both. Such rooms and spaces that are also part of a dwelling unit are not *sleeping units*.

SMALL ELECTRIC MOTOR. A general purpose, alternating current, single speed induction motor.

SOLAR HEAT GAIN COEFFICIENT (SHGC). The ratio of the solar heat gain entering the space through the fenestration assembly to the incident solar radiation. Solar heat gain includes directly transmitted solar heat and absorbed solar radiation which is then reradiated, conducted or convected into the space.

STANDARD REFERENCE DESIGN. A version of the *proposed design* that meets the minimum requirements of this code and is used to determine the maximum annual energy use requirement for compliance based on total building performance.

STOREFRONT. A nonresidential system of doors and windows mulled as a composite fenestration structure that has been designed to resist heavy use. *Storefront* systems include, but are not limited to, exterior fenestration systems that span from the floor level or above to the ceiling of the same story on commercial buildings, with or without mulled windows and doors.

THERMOSTAT. An automatic control device used to maintain temperature at a fixed or adjustable set point.

TIME SWITCH CONTROL. An automatic control device or system that controls lighting or other loads, including switching off, based on time schedules.

U-FACTOR (THERMAL TRANSMITTANCE). The coefficient of heat transmission (air to air) through a building component or assembly, equal to the time rate of heat flow per unit area and unit temperature difference between the warm side and cold side air films ($Btu/h \cdot ft^2 \cdot °F$) [$W/(m^2 \cdot K)$].

VARIABLE REFRIGERANT FLOW SYSTEM. An engineered direct-expansion (DX) refrigerant system that incorporates a common condensing unit, at least one variable-capacity compressor, a distributed refrigerant piping network to multiple indoor fan heating and cooling units each capable of individual zone temperature control, through integral zone temperature control devices and a common communications network. Variable refrigerant flow utilizes three or more steps of control on common interconnecting piping.

[M] VENTILATION. The natural or mechanical process of supplying conditioned or unconditioned air to, or removing such air from, any space.

[M] VENTILATION AIR. That portion of supply air that comes from outside (outdoors) plus any recirculated air that has been treated to maintain the desired quality of air within a designated space.

VISIBLE TRANSMITTANCE [VT]. The ratio of visible light entering the space through the fenestration product assembly to the incident visible light. Visible transmittance includes the effects of glazing material and frame and is expressed as a number between 0 and 1.

WALK-IN COOLER. An enclosed storage space capable of being refrigerated to temperatures above 32°F (0°C) and less than 55°F (12.8°C) that can be walked into, has a ceiling height of not less than 7 feet (2134 mm) and has a total chilled storage area of less than 3,000 square feet (279 m²).

WALK-IN FREEZER. An enclosed storage space capable of being refrigerated to temperatures at or below 32°F (0°C) that can be walked into, has a ceiling height of not less than 7

DEFINITIONS

feet (2134 mm) and has a total chilled storage area of less than 3,000 square feet (279 m^2).

WALL, ABOVE-GRADE. A wall associated with the *building thermal envelope* that is more than 15 percent above grade and is on the exterior of the building or any wall that is associated with the *building thermal envelope* that is not on the exterior of the building.

WALL, BELOW-GRADE. A wall associated with the basement or first story of the building that is part of the *building thermal envelope*, is not less than 85 percent below grade and is on the exterior of the building.

WATER HEATER. Any heating appliance or equipment that heats potable water and supplies such water to the potable hot water distribution system.

ZONE. A space or group of spaces within a building with heating or cooling requirements that are sufficiently similar so that desired conditions can be maintained throughout using a single controlling device.

CHAPTER 3 [CE]
GENERAL REQUIREMENTS

SECTION C301
CLIMATE ZONES

C301.1 General. *Climate zones* from Figure C301.1 or Table C301.1 shall be used in determining the applicable requirements from Chapter 4. Locations not in Table C301.1 (outside the United States) shall be assigned a *climate zone* based on Section C301.3.

C301.2 Warm humid counties. Warm humid counties are identified in Table C301.1 by an asterisk.

C301.3 International climate zones. The *climate zone* for any location outside the United States shall be determined by applying Table C301.3(1) and then Table C301.3(2).

C301.4 Tropical climate zone. The tropical *climate zone* shall be defined as:

1. Hawaii, Puerto Rico, Guam, American Samoa, U.S. Virgin Islands, Commonwealth of Northern Mariana Islands; and

2. Islands in the area between the Tropic of Cancer and the Tropic of Capricorn.

GENERAL REQUIREMENTS

FIGURE C301.1
CLIMATE ZONES

TABLE C301.1
CLIMATE ZONES, MOISTURE REGIMES, AND WARM-HUMID DESIGNATIONS BY STATE, COUNTY AND TERRITORY

Key: A – Moist, B – Dry, C – Marine. Absence of moisture designation indicates moisture regime is irrelevant. Asterisk (*) indicates a warm-humid location.

US STATES

ALABAMA

3A Autauga*
2A Baldwin*
3A Barbour*
3A Bibb
3A Blount
3A Bullock*
3A Butler*
3A Calhoun
3A Chambers
3A Cherokee
3A Chilton
3A Choctaw*
3A Clarke*
3A Clay
3A Cleburne
3A Coffee*
3A Colbert
3A Conecuh*
3A Coosa
3A Covington*
3A Crenshaw*
3A Cullman
3A Dale*
3A Dallas*
3A DeKalb
3A Elmore*
3A Escambia*
3A Etowah
3A Fayette
3A Franklin
3A Geneva*
3A Greene
3A Hale
3A Henry*
3A Houston*
3A Jackson
3A Jefferson
3A Lamar
3A Lauderdale
3A Lawrence

3A Lee
3A Limestone
3A Lowndes*
3A Macon*
3A Madison
3A Marengo*
3A Marion
3A Marshall
2A Mobile*
3A Monroe*
3A Montgomery*
3A Morgan
3A Perry*
3A Pickens
3A Pike*
3A Randolph
3A Russell*
3A Shelby
3A St. Clair
3A Sumter
3A Talladega
3A Tallapoosa
3A Tuscaloosa
3A Walker
3A Washington*
3A Wilcox*
3A Winston

ALASKA

7 Aleutians East
7 Aleutians West
7 Anchorage
8 Bethel
7 Bristol Bay
7 Denali
8 Dillingham
8 Fairbanks North Star
7 Haines
7 Juneau
7 Kenai Peninsula
7 Ketchikan Gateway

7 Kodiak Island
7 Lake and Peninsula
7 Matanuska-Susitna
8 Nome
8 North Slope
8 Northwest Arctic
7 Prince of Wales-Outer Ketchikan
7 Sitka
7 Skagway-Hoonah-Angoon
8 Southeast Fairbanks
7 Valdez-Cordova
8 Wade Hampton
7 Wrangell-Petersburg
7 Yakutat
8 Yukon-Koyukuk

ARIZONA

5B Apache
3B Cochise
5B Coconino
4B Gila
3B Graham
3B Greenlee
2B La Paz
2B Maricopa
3B Mohave
5B Navajo
2B Pima
2B Pinal
3B Santa Cruz
4B Yavapai
2B Yuma

ARKANSAS

3A Arkansas
3A Ashley
4A Baxter
4A Benton
4A Boone
3A Bradley

3A Calhoun
4A Carroll
3A Chicot
3A Clark
3A Clay
3A Cleburne
3A Cleveland
3A Columbia*
3A Conway
3A Craighead
3A Crawford
3A Crittenden
3A Cross
3A Dallas
3A Desha
3A Drew
3A Faulkner
3A Franklin
4A Fulton
3A Garland
3A Grant
3A Greene
3A Hempstead*
3A Hot Spring
3A Howard
3A Independence
4A Izard
3A Jackson
3A Jefferson
3A Johnson
3A Lafayette*
3A Lawrence
3A Lee
3A Lincoln
3A Little River*
3A Logan
3A Lonoke
4A Madison
4A Marion
3A Miller*
3A Mississippi

3A Monroe
3A Montgomery
3A Nevada
4A Newton
3A Ouachita
3A Perry
3A Phillips
3A Pike
3A Poinsett
3A Polk
3A Pope
3A Prairie
3A Pulaski
3A Randolph
3A Saline
3A Scott
4A Searcy
3A Sebastian
3A Sevier*
3A Sharp
3A St. Francis
4A Stone
3A Union*
3A Van Buren
4A Washington
3A White
3A Woodruff
3A Yell

CALIFORNIA

3C Alameda
6B Alpine
4B Amador
3B Butte
4B Calaveras
3B Colusa
3B Contra Costa
4C Del Norte
4B El Dorado
3B Fresno
3B Glenn

(continued)

GENERAL REQUIREMENTS

TABLE C301.1—continued
CLIMATE ZONES, MOISTURE REGIMES, AND WARM-HUMID DESIGNATIONS BY STATE, COUNTY AND TERRITORY

4C Humboldt	3B Yuba	5B Montrose	2A Escambia*	2A Taylor*
2B Imperial	**COLORADO**	5B Morgan	2A Flagler*	2A Union*
4B Inyo	5B Adams	4B Otero	2A Franklin*	2A Volusia*
3B Kern	6B Alamosa	6B Ouray	2A Gadsden*	2A Wakulla*
3B Kings	5B Arapahoe	7 Park	2A Gilchrist*	2A Walton*
4B Lake	6B Archuleta	5B Phillips	2A Glades*	2A Washington*
5B Lassen	4B Baca	7 Pitkin	2A Gulf*	**GEORGIA**
3B Los Angeles	5B Bent	5B Prowers	2A Hamilton*	2A Appling*
3B Madera	5B Boulder	5B Pueblo	2A Hardee*	2A Atkinson*
3C Marin	5B Broomfield	6B Rio Blanco	2A Hendry*	2A Bacon*
4B Mariposa	6B Chaffee	7 Rio Grande	2A Hernando*	2A Baker*
3C Mendocino	5B Cheyenne	7 Routt	2A Highlands*	3A Baldwin
3B Merced	7 Clear Creek	6B Saguache	2A Hillsborough*	4A Banks
5B Modoc	6B Conejos	7 San Juan	2A Holmes*	3A Barrow
6B Mono	6B Costilla	6B San Miguel	2A Indian River*	3A Bartow
3C Monterey	5B Crowley	5B Sedgwick	2A Jackson*	3A Ben Hill*
3C Napa	6B Custer	7 Summit	2A Jefferson*	2A Berrien*
5B Nevada	5B Delta	5B Teller	2A Lafayette*	3A Bibb
3B Orange	5B Denver	5B Washington	2A Lake*	3A Bleckley*
3B Placer	6B Dolores	5B Weld	2A Lee*	2A Brantley*
5B Plumas	5B Douglas	5B Yuma	2A Leon*	2A Brooks*
3B Riverside	6B Eagle	**CONNECTICUT**	2A Levy*	2A Bryan*
3B Sacramento	5B Elbert	5A (all)	2A Liberty*	3A Bulloch*
3C San Benito	5B El Paso	**DELAWARE**	2A Madison*	3A Burke
3B San Bernardino	5B Fremont	4A (all)	2A Manatee*	3A Butts
3B San Diego	5B Garfield	**DISTRICT OF COLUMBIA**	2A Marion*	3A Calhoun*
3C San Francisco	5B Gilpin		2A Martin*	2A Camden*
3B San Joaquin	7 Grand	4A (all)	1A Miami-Dade*	3A Candler*
3C San Luis Obispo	7 Gunnison	**FLORIDA**	1A Monroe*	3A Carroll
3C San Mateo	7 Hinsdale	2A Alachua*	2A Nassau*	4A Catoosa
3C Santa Barbara	5B Huerfano	2A Baker*	2A Okaloosa*	2A Charlton*
3C Santa Clara	7 Jackson	2A Bay*	2A Okeechobee*	2A Chatham*
3C Santa Cruz	5B Jefferson	2A Bradford*	2A Orange*	3A Chattahoochee*
3B Shasta	5B Kiowa	2A Brevard*	2A Osceola*	4A Chattooga
5B Sierra	5B Kit Carson	1A Broward*	2A Palm Beach*	3A Cherokee
5B Siskiyou	7 Lake	2A Calhoun*	2A Pasco*	3A Clarke
3B Solano	5B La Plata	2A Charlotte*	2A Pinellas*	3A Clay*
3C Sonoma	5B Larimer	2A Citrus*	2A Polk*	3A Clayton
3B Stanislaus	4B Las Animas	2A Clay*	2A Putnam*	2A Clinch*
3B Sutter	5B Lincoln	2A Collier*	2A Santa Rosa*	3A Cobb
3B Tehama	5B Logan	2A Columbia*	2A Sarasota*	3A Coffee*
4B Trinity	5B Mesa	2A DeSoto*	2A Seminole*	2A Colquitt*
3B Tulare	7 Mineral	2A Dixie*	2A St. Johns*	3A Columbia
4B Tuolumne	6B Moffat	2A Duval*	2A St. Lucie*	2A Cook*
3C Ventura	5B Montezuma		2A Sumter*	3A Coweta
3B Yolo			2A Suwannee*	

(continued)

GENERAL REQUIREMENTS

**TABLE C301.1—continued
CLIMATE ZONES, MOISTURE REGIMES, AND WARM-HUMID
DESIGNATIONS BY STATE, COUNTY AND TERRITORY**

3A Crawford	2A Lanier*	3A Taylor*	5B Cassia	4A Crawford
3A Crisp*	3A Laurens*	3A Telfair*	6B Clark	5A Cumberland
4A Dade	3A Lee*	3A Terrell*	5B Clearwater	5A DeKalb
4A Dawson	2A Liberty*	2A Thomas*	6B Custer	5A De Witt
2A Decatur*	3A Lincoln	3A Tift*	5B Elmore	5A Douglas
3A DeKalb	2A Long*	2A Toombs*	6B Franklin	5A DuPage
3A Dodge*	2A Lowndes*	4A Towns	6B Fremont	5A Edgar
3A Dooly*	4A Lumpkin	3A Treutlen*	5B Gem	4A Edwards
3A Dougherty*	3A Macon*	3A Troup	5B Gooding	4A Effingham
3A Douglas	3A Madison	3A Turner*	5B Idaho	4A Fayette
3A Early*	3A Marion*	3A Twiggs*	6B Jefferson	5A Ford
2A Echols*	3A McDuffie	4A Union	5B Jerome	4A Franklin
2A Effingham*	2A McIntosh*	3A Upson	5B Kootenai	5A Fulton
3A Elbert	3A Meriwether	4A Walker	5B Latah	4A Gallatin
3A Emanuel*	2A Miller*	3A Walton	6B Lemhi	5A Greene
2A Evans*	2A Mitchell*	2A Ware*	5B Lewis	5A Grundy
4A Fannin	3A Monroe	3A Warren	5B Lincoln	4A Hamilton
3A Fayette	3A Montgomery*	3A Washington	6B Madison	5A Hancock
4A Floyd	3A Morgan	2A Wayne*	5B Minidoka	4A Hardin
3A Forsyth	4A Murray	3A Webster*	5B Nez Perce	5A Henderson
4A Franklin	3A Muscogee	3A Wheeler*	6B Oneida	5A Henry
3A Fulton	3A Newton	4A White	5B Owyhee	5A Iroquois
4A Gilmer	3A Oconee	4A Whitfield	5B Payette	4A Jackson
3A Glascock	3A Oglethorpe	3A Wilcox*	5B Power	4A Jasper
2A Glynn*	3A Paulding	3A Wilkes	5B Shoshone	4A Jefferson
4A Gordon	3A Peach*	3A Wilkinson	6B Teton	5A Jersey
2A Grady*	4A Pickens	3A Worth*	5B Twin Falls	5A Jo Daviess
3A Greene	2A Pierce*	**HAWAII**	6B Valley	4A Johnson
3A Gwinnett	3A Pike	1A (all)*	5B Washington	5A Kane
4A Habersham	3A Polk	**IDAHO**	**ILLINOIS**	5A Kankakee
4A Hall	3A Pulaski*	5B Ada	5A Adams	5A Kendall
3A Hancock	3A Putnam	6B Adams	4A Alexander	5A Knox
3A Haralson	3A Quitman*	6B Bannock	4A Bond	5A Lake
3A Harris	4A Rabun	6B Bear Lake	5A Boone	5A La Salle
3A Hart	3A Randolph*	5B Benewah	5A Brown	4A Lawrence
3A Heard	3A Richmond	6B Bingham	5A Bureau	5A Lee
3A Henry	3A Rockdale	6B Blaine	5A Calhoun	5A Livingston
3A Houston*	3A Schley*	6B Boise	5A Carroll	5A Logan
3A Irwin*	3A Screven*	6B Bonner	5A Cass	5A Macon
3A Jackson	2A Seminole*	6B Bonneville	5A Champaign	4A Macoupin
3A Jasper	3A Spalding	6B Boundary	4A Christian	4A Madison
2A Jeff Davis*	4A Stephens	6B Butte	5A Clark	4A Marion
3A Jefferson	3A Stewart*	6B Camas	4A Clay	5A Marshall
3A Jenkins*	3A Sumter*	5B Canyon	4A Clinton	5A Mason
3A Johnson*	3A Talbot	6B Caribou	5A Coles	4A Massac
3A Jones	3A Taliaferro		5A Cook	5A McDonough
3A Lamar	2A Tattnall*			5A McHenry

(continued)

GENERAL REQUIREMENTS

**TABLE C301.1—continued
CLIMATE ZONES, MOISTURE REGIMES, AND WARM-HUMID
DESIGNATIONS BY STATE, COUNTY AND TERRITORY**

5A McLean	5A Boone	5A Miami	5A Appanoose	5A Jasper
5A Menard	4A Brown	4A Monroe	5A Audubon	5A Jefferson
5A Mercer	5A Carroll	5A Montgomery	5A Benton	5A Johnson
4A Monroe	5A Cass	5A Morgan	6A Black Hawk	5A Jones
4A Montgomery	4A Clark	5A Newton	5A Boone	5A Keokuk
5A Morgan	5A Clay	5A Noble	6A Bremer	6A Kossuth
5A Moultrie	5A Clinton	4A Ohio	6A Buchanan	5A Lee
5A Ogle	4A Crawford	4A Orange	6A Buena Vista	5A Linn
5A Peoria	4A Daviess	5A Owen	6A Butler	5A Louisa
4A Perry	4A Dearborn	5A Parke	6A Calhoun	5A Lucas
5A Piatt	5A Decatur	4A Perry	5A Carroll	6A Lyon
5A Pike	5A De Kalb	4A Pike	5A Cass	5A Madison
4A Pope	5A Delaware	5A Porter	6A Cedar	5A Mahaska
4A Pulaski	4A Dubois	4A Posey	6A Cerro Gordo	5A Marion
5A Putnam	5A Elkhart	5A Pulaski	6A Cherokee	5A Marshall
4A Randolph	5A Fayette	5A Putnam	6A Chickasaw	5A Mills
4A Richland	4A Floyd	5A Randolph	5A Clarke	6A Mitchell
5A Rock Island	5A Fountain	4A Ripley	6A Clay	5A Monona
4A Saline	5A Franklin	5A Rush	6A Clayton	5A Monroe
5A Sangamon	5A Fulton	4A Scott	5A Clinton	5A Montgomery
5A Schuyler	4A Gibson	5A Shelby	5A Crawford	5A Muscatine
5A Scott	5A Grant	4A Spencer	5A Dallas	6A O'Brien
4A Shelby	4A Greene	5A Starke	5A Davis	6A Osceola
5A Stark	5A Hamilton	5A Steuben	5A Decatur	5A Page
4A St. Clair	5A Hancock	5A St. Joseph	6A Delaware	6A Palo Alto
5A Stephenson	4A Harrison	4A Sullivan	5A Des Moines	6A Plymouth
5A Tazewell	5A Hendricks	4A Switzerland	6A Dickinson	6A Pocahontas
4A Union	5A Henry	5A Tippecanoe	5A Dubuque	5A Polk
5A Vermilion	5A Howard	5A Tipton	6A Emmet	5A Pottawattamie
4A Wabash	5A Huntington	5A Union	6A Fayette	5A Poweshiek
5A Warren	4A Jackson	4A Vanderburgh	6A Floyd	5A Ringgold
4A Washington	5A Jasper	5A Vermillion	6A Franklin	6A Sac
4A Wayne	5A Jay	5A Vigo	5A Fremont	5A Scott
4A White	4A Jefferson	5A Wabash	5A Greene	5A Shelby
5A Whiteside	4A Jennings	5A Warren	6A Grundy	6A Sioux
5A Will	5A Johnson	4A Warrick	5A Guthrie	5A Story
4A Williamson	4A Knox	4A Washington	6A Hamilton	5A Tama
5A Winnebago	5A Kosciusko	5A Wayne	6A Hancock	5A Taylor
5A Woodford	5A LaGrange	5A Wells	6A Hardin	5A Union
INDIANA	5A Lake	5A White	5A Harrison	5A Van Buren
	5A LaPorte	5A Whitley	5A Henry	5A Wapello
5A Adams	4A Lawrence	**IOWA**	6A Howard	5A Warren
5A Allen	5A Madison		6A Humboldt	5A Washington
5A Bartholomew	5A Marion	5A Adair	6A Ida	5A Wayne
5A Benton	5A Marshall	5A Adams	5A Iowa	6A Webster
5A Blackford	4A Martin	6A Allamakee	5A Jackson	6A Winnebago

(continued)

TABLE C301.1—continued
CLIMATE ZONES, MOISTURE REGIMES, AND WARM-HUMID DESIGNATIONS BY STATE, COUNTY AND TERRITORY

6A Winneshiek
5A Woodbury
6A Worth
6A Wright

KANSAS

4A Allen
4A Anderson
4A Atchison
4A Barber
4A Barton
4A Bourbon
4A Brown
4A Butler
4A Chase
4A Chautauqua
4A Cherokee
5A Cheyenne
4A Clark
4A Clay
5A Cloud
4A Coffey
4A Comanche
4A Cowley
4A Crawford
5A Decatur
4A Dickinson
4A Doniphan
4A Douglas
4A Edwards
4A Elk
5A Ellis
4A Ellsworth
4A Finney
4A Ford
4A Franklin
4A Geary
5A Gove
5A Graham
4A Grant
4A Gray
5A Greeley
4A Greenwood
5A Hamilton
4A Harper
4A Harvey

4A Haskell
4A Hodgeman
4A Jackson
4A Jefferson
5A Jewell
4A Johnson
4A Kearny
4A Kingman
4A Kiowa
4A Labette
5A Lane
4A Leavenworth
4A Lincoln
4A Linn
5A Logan
4A Lyon
4A Marion
4A Marshall
4A McPherson
4A Meade
4A Miami
5A Mitchell
4A Montgomery
4A Morris
4A Morton
4A Nemaha
4A Neosho
5A Ness
5A Norton
4A Osage
5A Osborne
4A Ottawa
4A Pawnee
5A Phillips
4A Pottawatomie
4A Pratt
5A Rawlins
4A Reno
5A Republic
4A Rice
4A Riley
5A Rooks
4A Rush
4A Russell
4A Saline
5A Scott

4A Sedgwick
4A Seward
4A Shawnee
5A Sheridan
5A Sherman
5A Smith
4A Stafford
4A Stanton
4A Stevens
4A Sumner
5A Thomas
5A Trego
4A Wabaunsee
5A Wallace
4A Washington
5A Wichita
4A Wilson
4A Woodson
4A Wyandotte

KENTUCKY

4A (all)

LOUISIANA

2A Acadia*
2A Allen*
2A Ascension*
2A Assumption*
2A Avoyelles*
2A Beauregard*
3A Bienville*
3A Bossier*
3A Caddo*
2A Calcasieu*
3A Caldwell*
2A Cameron*
3A Catahoula*
3A Claiborne*
3A Concordia*
3A De Soto*
2A East Baton Rouge*
3A East Carroll
2A East Feliciana*
2A Evangeline*
3A Franklin*
3A Grant*
2A Iberia*

2A Iberville*
3A Jackson*
2A Jefferson*
2A Jefferson Davis*
2A Lafayette*
2A Lafourche*
3A La Salle*
3A Lincoln*
2A Livingston*
3A Madison*
3A Morehouse
3A Natchitoches*
2A Orleans*
3A Ouachita*
2A Plaquemines*
2A Pointe Coupee*
2A Rapides*
3A Red River*
3A Richland*
3A Sabine*
2A St. Bernard*
2A St. Charles*
2A St. Helena*
2A St. James*
2A St. John the Baptist*
2A St. Landry*
2A St. Martin*
2A St. Mary*
2A St. Tammany*
2A Tangipahoa*
3A Tensas*
2A Terrebonne*
3A Union*
2A Vermilion*
3A Vernon*
2A Washington*
3A Webster*
2A West Baton Rouge*
3A West Carroll
2A West Feliciana*
3A Winn*

MAINE

6A Androscoggin
7   Aroostook

6A Cumberland
6A Franklin
6A Hancock
6A Kennebec
6A Knox
6A Lincoln
6A Oxford
6A Penobscot
6A Piscataquis
6A Sagadahoc
6A Somerset
6A Waldo
6A Washington
6A York

MARYLAND

4A Allegany
4A Anne Arundel
4A Baltimore
4A Baltimore (city)
4A Calvert
4A Caroline
4A Carroll
4A Cecil
4A Charles
4A Dorchester
4A Frederick
5A Garrett
4A Harford
4A Howard
4A Kent
4A Montgomery
4A Prince George's
4A Queen Anne's
4A Somerset
4A St. Mary's
4A Talbot
4A Washington
4A Wicomico
4A Worcester

MASSACHSETTS

5A (all)

MICHIGAN

6A Alcona
6A Alger

(continued)

GENERAL REQUIREMENTS

TABLE C301.1—continued
CLIMATE ZONES, MOISTURE REGIMES, AND WARM-HUMID
DESIGNATIONS BY STATE, COUNTY AND TERRITORY

5A Allegan	7 Mackinac	6A Carver	7 Otter Tail	3A Clarke
6A Alpena	5A Macomb	7 Cass	7 Pennington	3A Clay
6A Antrim	6A Manistee	6A Chippewa	7 Pine	3A Coahoma
6A Arenac	6A Marquette	6A Chisago	6A Pipestone	3A Copiah*
7 Baraga	6A Mason	7 Clay	7 Polk	3A Covington*
5A Barry	6A Mecosta	7 Clearwater	6A Pope	3A DeSoto
5A Bay	6A Menominee	7 Cook	6A Ramsey	3A Forrest*
6A Benzie	5A Midland	6A Cottonwood	7 Red Lake	3A Franklin*
5A Berrien	6A Missaukee	7 Crow Wing	6A Redwood	3A George*
5A Branch	5A Monroe	6A Dakota	6A Renville	3A Greene*
5A Calhoun	5A Montcalm	6A Dodge	6A Rice	3A Grenada
5A Cass	6A Montmorency	6A Douglas	6A Rock	2A Hancock*
6A Charlevoix	5A Muskegon	6A Faribault	7 Roseau	2A Harrison*
6A Cheboygan	6A Newaygo	6A Fillmore	6A Scott	3A Hinds*
7 Chippewa	5A Oakland	6A Freeborn	6A Sherburne	3A Holmes
6A Clare	6A Oceana	6A Goodhue	6A Sibley	3A Humphreys
5A Clinton	6A Ogemaw	7 Grant	6A Stearns	3A Issaquena
6A Crawford	7 Ontonagon	6A Hennepin	6A Steele	3A Itawamba
6A Delta	6A Osceola	6A Houston	6A Stevens	2A Jackson*
6A Dickinson	6A Oscoda	7 Hubbard	7 St. Louis	3A Jasper
5A Eaton	6A Otsego	6A Isanti	6A Swift	3A Jefferson*
6A Emmet	5A Ottawa	7 Itasca	6A Todd	3A Jefferson Davis*
5A Genesee	6A Presque Isle	6A Jackson	6A Traverse	3A Jones*
6A Gladwin	6A Roscommon	7 Kanabec	6A Wabasha	3A Kemper
7 Gogebic	5A Saginaw	6A Kandiyohi	7 Wadena	3A Lafayette
6A Grand Traverse	6A Sanilac	7 Kittson	6A Waseca	3A Lamar*
5A Gratiot	7 Schoolcraft	7 Koochiching	6A Washington	3A Lauderdale
5A Hillsdale	5A Shiawassee	6A Lac qui Parle	6A Watonwan	3A Lawrence*
7 Houghton	5A St. Clair	7 Lake	7 Wilkin	3A Leake
6A Huron	5A St. Joseph	7 Lake of the Woods	6A Winona	3A Lee
5A Ingham	5A Tuscola	6A Le Sueur	6A Wright	3A Leflore
5A Ionia	5A Van Buren	6A Lincoln	6A Yellow Medicine	3A Lincoln*
6A Iosco	5A Washtenaw	6A Lyon		3A Lowndes
7 Iron	5A Wayne	7 Mahnomen	**MISSISSIPPI**	3A Madison
6A Isabella	6A Wexford	7 Marshall	3A Adams*	3A Marion*
5A Jackson	**MINNESOTA**	6A Martin	3A Alcorn	3A Marshall
5A Kalamazoo	7 Aitkin	6A McLeod	3A Amite*	3A Monroe
6A Kalkaska	6A Anoka	6A Meeker	3A Attala	3A Montgomery
5A Kent	7 Becker	7 Mille Lacs	3A Benton	3A Neshoba
7 Keweenaw	7 Beltrami	6A Morrison	3A Bolivar	3A Newton
6A Lake	6A Benton	6A Mower	3A Calhoun	3A Noxubee
5A Lapeer	6A Big Stone	6A Murray	3A Carroll	3A Oktibbeha
6A Leelanau	6A Blue Earth	6A Nicollet	3A Chickasaw	3A Panola
5A Lenawee	6A Brown	6A Nobles	3A Choctaw	2A Pearl River*
5A Livingston	7 Carlton	7 Norman	3A Claiborne*	3A Perry*
7 Luce		6A Olmsted		3A Pike*

(continued)

GENERAL REQUIREMENTS

TABLE C301.1—continued
CLIMATE ZONES, MOISTURE REGIMES, AND WARM-HUMID
DESIGNATIONS BY STATE, COUNTY AND TERRITORY

3A Pontotoc
3A Prentiss
3A Quitman
3A Rankin*
3A Scott
3A Sharkey
3A Simpson*
3A Smith*
2A Stone*
3A Sunflower
3A Tallahatchie
3A Tate
3A Tippah
3A Tishomingo
3A Tunica
3A Union
3A Walthall*
3A Warren*
3A Washington
3A Wayne*
3A Webster
3A Wilkinson*
3A Winston
3A Yalobusha
3A Yazoo

MISSOURI

5A Adair
5A Andrew
5A Atchison
4A Audrain
4A Barry
4A Barton
4A Bates
4A Benton
4A Bollinger
4A Boone
5A Buchanan
4A Butler
5A Caldwell
4A Callaway
4A Camden
4A Cape Girardeau
4A Carroll
4A Carter
4A Cass
4A Cedar

5A Chariton
4A Christian
5A Clark
4A Clay
5A Clinton
4A Cole
4A Cooper
4A Crawford
4A Dade
4A Dallas
5A Daviess
5A DeKalb
4A Dent
4A Douglas
4A Dunklin
4A Franklin
4A Gasconade
5A Gentry
4A Greene
5A Grundy
5A Harrison
4A Henry
4A Hickory
5A Holt
4A Howard
4A Howell
4A Iron
4A Jackson
4A Jasper
4A Jefferson
4A Johnson
5A Knox
4A Laclede
4A Lafayette
4A Lawrence
5A Lewis
4A Lincoln
5A Linn
5A Livingston
5A Macon
4A Madison
4A Maries
5A Marion
4A McDonald
5A Mercer
4A Miller

4A Mississippi
4A Moniteau
4A Monroe
4A Montgomery
4A Morgan
4A New Madrid
4A Newton
5A Nodaway
4A Oregon
4A Osage
4A Ozark
4A Pemiscot
4A Perry
4A Pettis
4A Phelps
5A Pike
4A Platte
4A Polk
4A Pulaski
5A Putnam
5A Ralls
4A Randolph
4A Ray
4A Reynolds
4A Ripley
4A Saline
5A Schuyler
5A Scotland
4A Scott
4A Shannon
5A Shelby
4A St. Charles
4A St. Clair
4A St. Francois
4A St. Louis
4A St. Louis (city)
4A Ste. Genevieve
4A Stoddard
4A Stone
5A Sullivan
4A Taney
4A Texas
4A Vernon
4A Warren
4A Washington
4A Wayne

4A Webster
5A Worth
4A Wright

MONTANA

6B (all)

NEBRASKA

5A (all)

NEVADA

5B Carson City (city)
5B Churchill
3B Clark
5B Douglas
5B Elko
5B Esmeralda
5B Eureka
5B Humboldt
5B Lander
5B Lincoln
5B Lyon
5B Mineral
5B Nye
5B Pershing
5B Storey
5B Washoe
5B White Pine

NEW HAMPSHIRE

6A Belknap
6A Carroll
5A Cheshire
6A Coos
6A Grafton
5A Hillsborough
6A Merrimack
5A Rockingham
5A Strafford
6A Sullivan

NEW JERSEY

4A Atlantic
5A Bergen
4A Burlington
4A Camden
4A Cape May

4A Cumberland
4A Essex
4A Gloucester
4A Hudson
5A Hunterdon
5A Mercer
4A Middlesex
4A Monmouth
5A Morris
4A Ocean
5A Passaic
4A Salem
5A Somerset
5A Sussex
4A Union
5A Warren

NEW MEXICO

4B Bernalillo
5B Catron
3B Chaves
4B Cibola
5B Colfax
4B Curry
4B DeBaca
3B Doña Ana
3B Eddy
4B Grant
4B Guadalupe
5B Harding
3B Hidalgo
3B Lea
4B Lincoln
5B Los Alamos
3B Luna
5B McKinley
5B Mora
3B Otero
4B Quay
5B Rio Arriba
4B Roosevelt
5B Sandoval
5B San Juan
5B San Miguel
5B Santa Fe
4B Sierra
4B Socorro

(continued)

GENERAL REQUIREMENTS

TABLE C301.1—continued
CLIMATE ZONES, MOISTURE REGIMES, AND WARM-HUMID
DESIGNATIONS BY STATE, COUNTY AND TERRITORY

5B Taos	4A Queens	4A Clay	4A Orange	7 Divide
5B Torrance	5A Rensselaer	4A Cleveland	3A Pamlico	6A Dunn
4B Union	4A Richmond	3A Columbus*	3A Pasquotank	7 Eddy
4B Valencia	5A Rockland	3A Craven	3A Pender*	6A Emmons
NEW YORK	5A Saratoga	3A Cumberland	3A Perquimans	7 Foster
5A Albany	5A Schenectady	3A Currituck	4A Person	6A Golden Valley
6A Allegany	6A Schoharie	3A Dare	3A Pitt	7 Grand Forks
4A Bronx	5A Schuyler	3A Davidson	4A Polk	6A Grant
6A Broome	5A Seneca	4A Davie	3A Randolph	7 Griggs
6A Cattaraugus	6A Steuben	3A Duplin	3A Richmond	6A Hettinger
5A Cayuga	6A St. Lawrence	4A Durham	3A Robeson	7 Kidder
5A Chautauqua	4A Suffolk	3A Edgecombe	4A Rockingham	6A LaMoure
5A Chemung	6A Sullivan	4A Forsyth	3A Rowan	6A Logan
6A Chenango	5A Tioga	4A Franklin	4A Rutherford	7 McHenry
6A Clinton	6A Tompkins	3A Gaston	3A Sampson	6A McIntosh
5A Columbia	6A Ulster	4A Gates	3A Scotland	6A McKenzie
5A Cortland	6A Warren	4A Graham	3A Stanly	7 McLean
6A Delaware	5A Washington	4A Granville	4A Stokes	6A Mercer
5A Dutchess	5A Wayne	3A Greene	4A Surry	6A Morton
5A Erie	4A Westchester	4A Guilford	4A Swain	7 Mountrail
6A Essex	6A Wyoming	4A Halifax	4A Transylvania	7 Nelson
6A Franklin	5A Yates	4A Harnett	3A Tyrrell	6A Oliver
6A Fulton	**NORTH CAROLINA**	4A Haywood	3A Union	7 Pembina
5A Genesee		4A Henderson	4A Vance	7 Pierce
5A Greene		4A Hertford	4A Wake	7 Ramsey
6A Hamilton	4A Alamance	3A Hoke	4A Warren	6A Ransom
6A Herkimer	4A Alexander	3A Hyde	3A Washington	7 Renville
6A Jefferson	5A Alleghany	4A Iredell	5A Watauga	6A Richland
4A Kings	3A Anson	4A Jackson	3A Wayne	7 Rolette
6A Lewis	5A Ashe	3A Johnston	4A Wilkes	6A Sargent
5A Livingston	5A Avery	3A Jones	3A Wilson	7 Sheridan
6A Madison	3A Beaufort	4A Lee	4A Yadkin	6A Sioux
5A Monroe	4A Bertie	3A Lenoir	5A Yancey	6A Slope
6A Montgomery	3A Bladen	4A Lincoln	**NORTH DAKOTA**	6A Stark
4A Nassau	3A Brunswick*	4A Macon		7 Steele
4A New York	4A Buncombe	4A Madison	6A Adams	7 Stutsman
5A Niagara	4A Burke	3A Martin	7 Barnes	7 Towner
6A Oneida	3A Cabarrus	4A McDowell	7 Benson	7 Traill
5A Onondaga	4A Caldwell	3A Mecklenburg	6A Billings	7 Walsh
5A Ontario	3A Camden	5A Mitchell	7 Bottineau	7 Ward
5A Orange	3A Carteret*	3A Montgomery	6A Bowman	7 Wells
5A Orleans	4A Caswell	3A Moore	7 Burke	7 Williams
5A Oswego	4A Catawba	4A Nash	6A Burleigh	**OHIO**
6A Otsego	4A Chatham	3A New Hanover*	7 Cass	
5A Putnam	4A Cherokee	4A Northampton	7 Cavalier	4A Adams
	3A Chowan	3A Onslow*	6A Dickey	5A Allen

(continued)

TABLE C301.1—continued
CLIMATE ZONES, MOISTURE REGIMES, AND WARM-HUMID DESIGNATIONS BY STATE, COUNTY AND TERRITORY

5A Ashland	5A Mahoning	3A Bryan	3A Okfuskee	4C Linn
5A Ashtabula	5A Marion	3A Caddo	3A Oklahoma	5B Malheur
5A Athens	5A Medina	3A Canadian	3A Okmulgee	4C Marion
5A Auglaize	5A Meigs	3A Carter	3A Osage	5B Morrow
5A Belmont	5A Mercer	3A Cherokee	3A Ottawa	4C Multnomah
4A Brown	5A Miami	3A Choctaw	3A Pawnee	4C Polk
5A Butler	5A Monroe	4B Cimarron	3A Payne	5B Sherman
5A Carroll	5A Montgomery	3A Cleveland	3A Pittsburg	4C Tillamook
5A Champaign	5A Morgan	3A Coal	3A Pontotoc	5B Umatilla
5A Clark	5A Morrow	3A Comanche	3A Pottawatomie	5B Union
4A Clermont	5A Muskingum	3A Cotton	3A Pushmataha	5B Wallowa
5A Clinton	5A Noble	3A Craig	3A Roger Mills	5B Wasco
5A Columbiana	5A Ottawa	3A Creek	3A Rogers	4C Washington
5A Coshocton	5A Paulding	3A Custer	3A Seminole	5B Wheeler
5A Crawford	5A Perry	3A Delaware	3A Sequoyah	4C Yamhill
5A Cuyahoga	5A Pickaway	3A Dewey	3A Stephens	**PENNSYLVANIA**
5A Darke	4A Pike	3A Ellis	4B Texas	5A Adams
5A Defiance	5A Portage	3A Garfield	3A Tillman	5A Allegheny
5A Delaware	5A Preble	3A Garvin	3A Tulsa	5A Armstrong
5A Erie	5A Putnam	3A Grady	3A Wagoner	5A Beaver
5A Fairfield	5A Richland	3A Grant	3A Washington	5A Bedford
5A Fayette	5A Ross	3A Greer	3A Washita	5A Berks
5A Franklin	5A Sandusky	3A Harmon	3A Woods	5A Blair
5A Fulton	4A Scioto	3A Harper	3A Woodward	5A Bradford
4A Gallia	5A Seneca	3A Haskell	**OREGON**	4A Bucks
5A Geauga	5A Shelby	3A Hughes	5B Baker	5A Butler
5A Greene	5A Stark	3A Jackson	4C Benton	5A Cambria
5A Guernsey	5A Summit	3A Jefferson	4C Clackamas	6A Cameron
4A Hamilton	5A Trumbull	3A Johnston	4C Clatsop	5A Carbon
5A Hancock	5A Tuscarawas	3A Kay	4C Columbia	5A Centre
5A Hardin	5A Union	3A Kingfisher	4C Coos	4A Chester
5A Harrison	5A Van Wert	3A Kiowa	5B Crook	5A Clarion
5A Henry	5A Vinton	3A Latimer	4C Curry	6A Clearfield
5A Highland	5A Warren	3A Le Flore	5B Deschutes	5A Clinton
5A Hocking	4A Washington	3A Lincoln	4C Douglas	5A Columbia
5A Holmes	5A Wayne	3A Logan	5B Gilliam	5A Crawford
5A Huron	5A Williams	3A Love	5B Grant	5A Cumberland
5A Jackson	5A Wood	3A Major	5B Harney	5A Dauphin
5A Jefferson	5A Wyandot	3A Marshall	5B Hood River	4A Delaware
5A Knox	**OKLAHOMA**	3A Mayes	4C Jackson	6A Elk
5A Lake	3A Adair	3A McClain	5B Jefferson	5A Erie
4A Lawrence	3A Alfalfa	3A McCurtain	4C Josephine	5A Fayette
5A Licking	3A Atoka	3A McIntosh	5B Klamath	5A Forest
5A Logan	4B Beaver	3A Murray	5B Lake	5A Franklin
5A Lorain	3A Beckham	3A Muskogee	4C Lane	5A Fulton
5A Lucas	3A Blaine	3A Noble	4C Lincoln	5A Greene
5A Madison		3A Nowata		

(continued)

GENERAL REQUIREMENTS

TABLE C301.1—continued
CLIMATE ZONES, MOISTURE REGIMES, AND WARM-HUMID DESIGNATIONS BY STATE, COUNTY AND TERRITORY

5A Huntingdon	3A Bamberg*	5A Bennett	6A Minnehaha	4A Gibson
5A Indiana	3A Barnwell*	5A Bon Homme	6A Moody	4A Giles
5A Jefferson	3A Beaufort*	6A Brookings	6A Pennington	4A Grainger
5A Juniata	3A Berkeley*	6A Brown	6A Perkins	4A Greene
5A Lackawanna	3A Calhoun	6A Brule	6A Potter	4A Grundy
5A Lancaster	3A Charleston*	6A Buffalo	6A Roberts	4A Hamblen
5A Lawrence	3A Cherokee	6A Butte	6A Sanborn	4A Hamilton
5A Lebanon	3A Chester	6A Campbell	6A Shannon	4A Hancock
5A Lehigh	3A Chesterfield	5A Charles Mix	6A Spink	3A Hardeman
5A Luzerne	3A Clarendon	6A Clark	6A Stanley	3A Hardin
5A Lycoming	3A Colleton*	5A Clay	6A Sully	4A Hawkins
6A McKean	3A Darlington	6A Codington	5A Todd	3A Haywood
5A Mercer	3A Dillon	6A Corson	5A Tripp	3A Henderson
5A Mifflin	3A Dorchester*	6A Custer	6A Turner	4A Henry
5A Monroe	3A Edgefield	6A Davison	5A Union	4A Hickman
4A Montgomery	3A Fairfield	6A Day	6A Walworth	4A Houston
5A Montour	3A Florence	6A Deuel	5A Yankton	4A Humphreys
5A Northampton	3A Georgetown*	6A Dewey	6A Ziebach	4A Jackson
5A Northumberland	3A Greenville	5A Douglas	**TENNESSEE**	4A Jefferson
5A Perry	3A Greenwood	6A Edmunds	4A Anderson	4A Johnson
4A Philadelphia	3A Hampton*	6A Fall River	4A Bedford	4A Knox
5A Pike	3A Horry*	6A Faulk	4A Benton	3A Lake
6A Potter	3A Jasper*	6A Grant	4A Bledsoe	3A Lauderdale
5A Schuylkill	3A Kershaw	5A Gregory	4A Blount	4A Lawrence
5A Snyder	3A Lancaster	6A Haakon	4A Bradley	4A Lewis
5A Somerset	3A Laurens	6A Hamlin	4A Campbell	4A Lincoln
5A Sullivan	3A Lee	6A Hand	4A Cannon	4A Loudon
6A Susquehanna	3A Lexington	6A Hanson	4A Carroll	4A Macon
6A Tioga	3A Marion	6A Harding	4A Carter	3A Madison
5A Union	3A Marlboro	6A Hughes	4A Cheatham	4A Marion
5A Venango	3A McCormick	5A Hutchinson	3A Chester	4A Marshall
5A Warren	3A Newberry	6A Hyde	4A Claiborne	4A Maury
5A Washington	3A Oconee	5A Jackson	4A Clay	4A McMinn
6A Wayne	3A Orangeburg	6A Jerauld	4A Cocke	3A McNairy
5A Westmoreland	3A Pickens	6A Jones	4A Coffee	4A Meigs
5A Wyoming	3A Richland	6A Kingsbury	3A Crockett	4A Monroe
4A York	3A Saluda	6A Lake	4A Cumberland	4A Montgomery
RHODE ISLAND	3A Spartanburg	6A Lawrence	4A Davidson	4A Moore
5A (all)	3A Sumter	6A Lincoln	4A Decatur	4A Morgan
SOUTH CAROLINA	3A Union	6A Lyman	4A DeKalb	4A Obion
	3A Williamsburg	6A Marshall	4A Dickson	4A Overton
	3A York	6A McCook	3A Dyer	4A Perry
3A Abbeville	**SOUTH DAKOTA**	6A McPherson	3A Fayette	4A Pickett
3A Aiken	6A Aurora	6A Meade	4A Fentress	4A Polk
3A Allendale*	6A Beadle	5A Mellette	4A Franklin	4A Putnam
3A Anderson		6A Miner		4A Rhea

(continued)

GENERAL REQUIREMENTS

TABLE C301.1—continued
CLIMATE ZONES, MOISTURE REGIMES, AND WARM-HUMID
DESIGNATIONS BY STATE, COUNTY AND TERRITORY

4A Roane	3B Brewster	3B Ector	3B Howard	3B McCulloch
4A Robertson	4B Briscoe	2B Edwards	3B Hudspeth	2A McLennan*
4A Rutherford	2A Brooks*	3A Ellis*	3A Hunt*	2A McMullen*
4A Scott	3A Brown*	3B El Paso	4B Hutchinson	2B Medina
4A Sequatchie	2A Burleson*	3A Erath*	3B Irion	3B Menard
4A Sevier	3A Burnet*	2A Falls*	3A Jack	3B Midland
3A Shelby	2A Caldwell*	3A Fannin	2A Jackson*	2A Milam*
4A Smith	2A Calhoun*	2A Fayette*	2A Jasper*	3A Mills*
4A Stewart	3B Callahan	3B Fisher	3B Jeff Davis	3B Mitchell
4A Sullivan	2A Cameron*	4B Floyd	2A Jefferson*	3A Montague
4A Sumner	3A Camp*	3B Foard	2A Jim Hogg*	2A Montgomery*
3A Tipton	4B Carson	2A Fort Bend*	2A Jim Wells*	4B Moore
4A Trousdale	3A Cass*	3A Franklin*	3A Johnson*	3A Morris*
4A Unicoi	4B Castro	2A Freestone*	3B Jones	3B Motley
4A Union	2A Chambers*	2B Frio	2A Karnes*	3A Nacogdoches*
4A Van Buren	2A Cherokee*	3B Gaines	3A Kaufman*	3A Navarro*
4A Warren	3B Childress	2A Galveston*	2A Kendall*	2A Newton*
4A Washington	3A Clay	3B Garza	2A Kenedy*	3B Nolan
4A Wayne	4B Cochran	3A Gillespie*	3B Kent	2A Nueces*
4A Weakley	3B Coke	3B Glasscock	3B Kerr	4B Ochiltree
4A White	3B Coleman	2A Goliad*	3B Kimble	4B Oldham
4A Williamson	3A Collin*	2A Gonzales*	3B King	2A Orange*
4A Wilson	3B Collingsworth	4B Gray	2B Kinney	3A Palo Pinto*
TEXAS	2A Colorado*	3A Grayson	2A Kleberg*	3A Panola*
2A Anderson*	2A Comal*	3A Gregg*	3B Knox	3A Parker*
3B Andrews	3A Comanche*	2A Grimes*	3A Lamar*	4B Parmer
2A Angelina*	3B Concho	2A Guadalupe*	4B Lamb	3B Pecos
2A Aransas*	3A Cooke	4B Hale	3A Lampasas*	2A Polk*
3A Archer	2A Coryell*	3B Hall	2B La Salle	4B Potter
4B Armstrong	3B Cottle	3A Hamilton*	2A Lavaca*	3B Presidio
2A Atascosa*	3B Crane	4B Hansford	2A Lee*	3A Rains*
2A Austin*	3B Crockett	3B Hardeman	2A Leon*	4B Randall
4B Bailey	3B Crosby	2A Hardin*	2A Liberty*	3B Reagan
2B Bandera	3B Culberson	2A Harris*	2A Limestone*	2B Real
2A Bastrop*	4B Dallam	3A Harrison*	4B Lipscomb	3A Red River*
3B Baylor	3A Dallas*	4B Hartley	2A Live Oak*	3B Reeves
2A Bee*	3B Dawson	3B Haskell	3A Llano*	2A Refugio*
2A Bell*	4B Deaf Smith	2A Hays*	3B Loving	4B Roberts
2A Bexar*	3A Delta	3B Hemphill	3B Lubbock	2A Robertson*
3A Blanco*	3A Denton*	3A Henderson*	3B Lynn	3A Rockwall*
3B Borden	2A DeWitt*	2A Hidalgo*	2A Madison*	3B Runnels
2A Bosque*	3B Dickens	2A Hill*	3A Marion*	3A Rusk*
3A Bowie*	2B Dimmit	4B Hockley	3B Martin	3A Sabine*
2A Brazoria*	4B Donley	3A Hood*	3B Mason	3A San Augustine*
2A Brazos*	2A Duval*	3A Hopkins*	2A Matagorda*	2A San Jacinto*
	3A Eastland	2A Houston*	2B Maverick	2A San Patricio*

(continued)

GENERAL REQUIREMENTS

TABLE C301.1—continued
CLIMATE ZONES, MOISTURE REGIMES, AND WARM-HUMID DESIGNATIONS BY STATE, COUNTY AND TERRITORY

3A San Saba*	3A Young	4C Clark	4A Gilmer	**WISCONSIN**
3B Schleicher	2B Zapata	5B Columbia	5A Grant	6A Adams
3B Scurry	2B Zavala	4C Cowlitz	5A Greenbrier	7　Ashland
3B Shackelford	**UTAH**	5B Douglas	5A Hampshire	6A Barron
3A Shelby*	5B Beaver	6B Ferry	5A Hancock	7　Bayfield
4B Sherman	6B Box Elder	5B Franklin	5A Hardy	6A Brown
3A Smith*	6B Cache	5B Garfield	5A Harrison	6A Buffalo
3A Somervell*	6B Carbon	5B Grant	4A Jackson	7　Burnett
2A Starr*	6B Daggett	4C Grays Harbor	4A Jefferson	6A Calumet
3A Stephens	5B Davis	4C Island	4A Kanawha	6A Chippewa
3B Sterling	6B Duchesne	4C Jefferson	5A Lewis	6A Clark
3B Stonewall	5B Emery	4C King	4A Lincoln	6A Columbia
3B Sutton	5B Garfield	4C Kitsap	4A Logan	6A Crawford
4B Swisher	5B Grand	5B Kittitas	5A Marion	6A Dane
3A Tarrant*	5B Iron	5B Klickitat	5A Marshall	6A Dodge
3B Taylor	5B Juab	4C Lewis	4A Mason	6A Door
3B Terrell	5B Kane	5B Lincoln	4A McDowell	7　Douglas
3B Terry	5B Millard	4C Mason	4A Mercer	6A Dunn
3B Throckmorton	6B Morgan	6B Okanogan	5A Mineral	6A Eau Claire
3A Titus*	5B Piute	4C Pacific	4A Mingo	7　Florence
3B Tom Green	6B Rich	6B Pend Oreille	5A Monongalia	6A Fond du Lac
2A Travis*	5B Salt Lake	4C Pierce	4A Monroe	7　Forest
2A Trinity*	5B San Juan	4C San Juan	4A Morgan	6A Grant
2A Tyler*	5B Sanpete	4C Skagit	5A Nicholas	6A Green
3A Upshur*	5B Sevier	5B Skamania	5A Ohio	6A Green Lake
3B Upton	6B Summit	4C Snohomish	5A Pendleton	6A Iowa
2B Uvalde	5B Tooele	5B Spokane	4A Pleasants	7　Iron
2B Val Verde	6B Uintah	6B Stevens	5A Pocahontas	6A Jackson
3A Van Zandt*	5B Utah	4C Thurston	5A Preston	6A Jefferson
2A Victoria*	6B Wasatch	4C Wahkiakum	4A Putnam	6A Juneau
2A Walker*	3B Washington	5B Walla Walla	5A Raleigh	6A Kenosha
2A Waller*	5B Wayne	4C Whatcom	5A Randolph	6A Kewaunee
3B Ward	5B Weber	5B Whitman	4A Ritchie	6A La Crosse
2A Washington*	**VERMONT**	5B Yakima	4A Roane	6A Lafayette
2B Webb	6A (all)	**WEST VIRGINIA**	5A Summers	7　Langlade
2A Wharton*	**VIRGINIA**	5A Barbour	5A Taylor	7　Lincoln
3B Wheeler	4A (all)	4A Berkeley	5A Tucker	6A Manitowoc
3A Wichita	**WASHINGTON**	4A Boone	4A Tyler	6A Marathon
3B Wilbarger	5B Adams	4A Braxton	5A Upshur	6A Marinette
2A Willacy*	5B Asotin	5A Brooke	4A Wayne	6A Marquette
2A Williamson*	5B Benton	4A Cabell	5A Webster	6A Menominee
2A Wilson*	5B Chelan	4A Calhoun	5A Wetzel	6A Milwaukee
3B Winkler	4C Clallam	4A Clay	4A Wirt	6A Monroe
3A Wise		5A Doddridge	4A Wood	6A Oconto
3A Wood*		5A Fayette	4A Wyoming	7　Oneida
4B Yoakum				6A Outagamie

(continued)

TABLE C301.1—continued
CLIMATE ZONES, MOISTURE REGIMES, AND WARM-HUMID DESIGNATIONS BY STATE, COUNTY AND TERRITORY

6A Ozaukee	7 Taylor	6B Big Horn	6B Sheridan	**NORTHERN MARIANA ISLANDS**
6A Pepin	6A Trempealeau	6B Campbell	7 Sublette	
6A Pierce	6A Vernon	6B Carbon	6B Sweetwater	
6A Polk	7 Vilas	6B Converse	7 Teton	1A (all)*
6A Portage	6A Walworth	6B Crook	6B Uinta	**PUERTO RICO**
7 Price	7 Washburn	6B Fremont	6B Washakie	
6A Racine	6A Washington	5B Goshen	6B Weston	1A (all)*
6A Richland	6A Waukesha	6B Hot Springs	**US TERRITORIES**	**VIRGIN ISLANDS**
6A Rock	6A Waupaca	6B Johnson		1A (all)*
6A Rusk	6A Waushara	6B Laramie	**AMERICAN SAMOA**	
6A Sauk	6A Winnebago	7 Lincoln		
7 Sawyer	6A Wood	6B Natrona	1A (all)*	
6A Shawano	**WYOMING**	6B Niobrara	**GUAM**	
6A Sheboygan		6B Park		
6A St. Croix	6B Albany	5B Platte	1A (all)*	

TABLE C301.3(1)
INTERNATIONAL CLIMATE ZONE DEFINITIONS

MAJOR CLIMATE TYPE DEFINITIONS
Marine (C) Definition—Locations meeting all four criteria: 1. Mean temperature of coldest month between -3°C (27°F) and 18°C (65°F).   2. Warmest month mean < 22°C (72°F).   3. At least four months with mean temperatures over 10°C (50°F).   4. Dry season in summer. The month with the heaviest precipitation in the cold season has at least three times as much precipitation as the month with the least precipitation in the rest of the year. The cold season is October through March in the Northern Hemisphere and April through September in the Southern Hemisphere.
Dry (B) Definition—Locations meeting the following criteria: Not marine and $P_{in} < 0.44 \times (TF - 19.5)$ $[P_{cm} < 2.0 \times (TC + 7)$ in SI units]   where: P_{in} = Annual precipitation in inches (cm) T = Annual mean temperature in °F (°C)
Moist (A) Definition—Locations that are not marine and not dry.
Warm-humid Definition—Moist (A) locations where either of the following wet-bulb temperature conditions shall occur during the warmest six consecutive months of the year: 1. 67°F (19.4°C) or higher for 3,000 or more hours; or   2. 73°F (22.8°C) or higher for 1,500 or more hours.

For SI: °C = [(°F)-32]/1.8, 1 inch = 2.54 cm.

GENERAL REQUIREMENTS

TABLE C301.3(2)
INTERNATIONAL CLIMATE ZONE DEFINITIONS

ZONE NUMBER	THERMAL CRITERIA	
	IP Units	SI Units
1	9000 < CDD50°F	5000 < CDD10°C
2	6300 < CDD50°F ≤ 9000	3500 < CDD10°C ≤ 5000
3A and 3B	4500 < CDD50°F ≤ 6300 AND HDD65°F ≤ 5400	2500 < CDD10°C ≤ 3500 AND HDD18°C ≤ 3000
4A and 4B	CDD50°F ≤ 4500 AND HDD65°F ≤ 5400	CDD10°C ≤ 2500 AND HDD18°C ≤ 3000
3C	HDD65°F ≤ 3600	HDD18°C ≤ 2000
4C	3600 < HDD65°F ≤ 5400	2000 < HDD18°C ≤ 3000
5	5400 < HDD65°F ≤ 7200	3000 < HDD18°C ≤ 4000
6	7200 < HDD65°F ≤ 9000	4000 < HDD18°C ≤ 5000
7	9000 < HDD65°F ≤ 12600	5000 < HDD18°C ≤ 7000
8	12600 < HDD65°F	7000 < HDD18°C

For SI: °C = [(°F)-32]/1.8.

SECTION C302
DESIGN CONDITIONS

C302.1 Interior design conditions. The interior design temperatures used for heating and cooling load calculations shall be a maximum of 72°F (22°C) for heating and minimum of 75°F (24°C) for cooling.

SECTION C303
MATERIALS, SYSTEMS AND EQUIPMENT

C303.1 Identification. Materials, systems and equipment shall be identified in a manner that will allow a determination of compliance with the applicable provisions of this code.

C303.1.1 Building thermal envelope insulation. An *R*-value identification mark shall be applied by the manufacturer to each piece of *building thermal envelope* insulation 12 inches (305 mm) or greater in width. Alternately, the insulation installers shall provide a certification listing the type, manufacturer and *R*-value of insulation installed in each element of the *building thermal envelope*. For blown or sprayed insulation (fiberglass and cellulose), the initial installed thickness, settled thickness, settled *R*-value, installed density, coverage area and number of bags installed shall be *listed* on the certification. For sprayed polyurethane foam (SPF) insulation, the installed thickness of the areas covered and *R*-value of installed thickness shall be *listed* on the certification. For insulated siding, the *R*-value shall be labeled on the product's package and shall be *listed* on the certification. The insulation installer shall sign, date and post the certification in a conspicuous location on the job site.

C303.1.1.1 Blown or sprayed roof/ceiling insulation. The thickness of blown-in or sprayed roof/ceiling insulation (fiberglass or cellulose) shall be written in inches (mm) on markers that are installed at least one for every 300 square feet (28 m^2) throughout the attic space. The markers shall be affixed to the trusses or joists and marked with the minimum initial installed thickness with numbers not less than 1 inch (25 mm) in height. Each marker shall face the attic access opening. Spray polyurethane foam thickness and installed *R*-value shall be *listed* on certification provided by the insulation installer.

C303.1.2 Insulation mark installation. Insulating materials shall be installed such that the manufacturer's *R*-value mark is readily observable upon inspection.

C303.1.3 Fenestration product rating. *U*-factors of fenestration products (windows, doors and skylights) shall be determined in accordance with NFRC 100.

> **Exception:** Where required, garage door *U*-factors shall be determined in accordance with either NFRC 100 or ANSI/DASMA 105.

U-factors shall be determined by an accredited, independent laboratory, and *labeled* and certified by the manufacturer.

Products lacking such a *labeled U*-factor shall be assigned a default *U*-factor from Table C303.1.3(1) or C303.1.3(2). The solar heat gain coefficient (SHGC) and *visible transmittance* (VT) of glazed fenestration products (windows, glazed doors and skylights) shall be determined in accordance with NFRC 200 by an accredited, independent laboratory, and *labeled* and certified by the manufacturer. Products lacking such a *labeled* SHGC or VT shall be assigned a default SHGC or VT from Table C303.1.3(3).

TABLE C303.1.3(1)
DEFAULT GLAZED FENESTRATION *U*-FACTORS

FRAME TYPE	SINGLE PANE	DOUBLE PANE	SKYLIGHT	
			Single	Double
Metal	1.20	0.80	2.00	1.30
Metal with Thermal Break	1.10	0.65	1.90	1.10
Nonmetal or Metal Clad	0.95	0.55	1.75	1.05
Glazed Block	0.60			

TABLE C303.1.3(2)
DEFAULT DOOR U-FACTORS

DOOR TYPE	U-FACTOR
Uninsulated Metal	1.20
Insulated Metal	0.60
Wood	0.50
Insulated, nonmetal edge, max 45% glazing, any glazing double pane	0.35

TABLE C303.1.3(3)
DEFAULT GLAZED FENESTRATION SHGC AND VT

	SINGLE GLAZED		DOUBLE GLAZED		GLAZED BLOCK
	Clear	Tinted	Clear	Tinted	
SHGC	0.8	0.7	0.7	0.6	0.6
VT	0.6	0.3	0.6	0.3	0.6

C303.1.4 Insulation product rating. The thermal resistance (R-value) of insulation shall be determined in accordance with the U.S. Federal Trade Commission R-value rule (CFR Title 16, Part 460) in units of h · ft^2 · °F/Btu at a mean temperature of 75°F (24°C).

C303.1.4.1 Insulated siding. The thermal resistance (R-value) of insulated siding shall be determined in accordance with ASTM C1363. Installation for testing shall be in accordance with the manufacturer's instructions.

C303.2 Installation. Materials, systems and equipment shall be installed in accordance with the manufacturer's instructions and the *International Building Code*.

C303.2.1 Protection of exposed foundation insulation. Insulation applied to the exterior of basement walls, crawlspace walls and the perimeter of slab-on-grade floors shall have a rigid, opaque and weather-resistant protective covering to prevent the degradation of the insulation's thermal performance. The protective covering shall cover the exposed exterior insulation and extend not less than 6 inches (153 mm) below grade.

C303.3 Maintenance information. Maintenance instructions shall be furnished for equipment and systems that require preventive maintenance. Required regular maintenance actions shall be clearly stated and incorporated on a readily accessible label. The label shall include the title or publication number for the operation and maintenance manual for that particular model and type of product.

CHAPTER 4 [CE]
COMMERCIAL ENERGY EFFICIENCY

SECTION C401
GENERAL

C401.1 Scope. The provisions in this chapter are applicable to commercial *buildings* and their *building sites*.

C401.2 Application. Commercial buildings shall comply with one of the following:

1. The requirements of ANSI/ASHRAE/IESNA 90.1.

2. The requirements of Sections C402 through C405. In addition, commercial buildings shall comply with Section C406 and tenant spaces shall comply with Section C406.1.1.

3. The requirements of Sections C402.5, C403.2, C404, C405.2, C405.3, C405.5, C405.6 and C407. The building energy cost shall be equal to or less than 85 percent of the standard reference design building.

C401.2.1 Application to replacement fenestration products. Where some or all of an existing *fenestration* unit is replaced with a new *fenestration* product, including sash and glazing, the replacement *fenestration* unit shall meet the applicable requirements for *U-factor* and *SHGC* in Table C402.4.

> **Exception:** An area-weighted average of the *U-factor* of replacement fenestration products being installed in the building for each fenestration product category listed in Table C402.4 shall be permitted to satisfy the *U-factor* requirements for each fenestration product category listed in Table C402.4. Individual fenestration products from different product categories listed in Table C402.3 shall not be combined in calculating the area-weighted average *U-factor*.

SECTION C402
BUILDING ENVELOPE REQUIREMENTS

C402.1 General (Prescriptive). Building thermal envelope assemblies for buildings that are intended to comply with the code on a prescriptive basis, in accordance with the compliance path described in Item 2 of Section C401.2, shall comply with the following:

1. The opaque portions of the building thermal envelope shall comply with the specific insulation requirements of Section C402.2 and the thermal requirements of either the *R*-value-based method of Section C402.1.3; the *U-*, *C-* and *F*-factor-based method of Section C402.1.4; or the component performance alternative of Section C402.1.5.

2. Roof solar reflectance and thermal emittance shall comply with Section C402.3.

3. Fenestration in building envelope assemblies shall comply with Section C402.4.

4. Air leakage of building envelope assemblies shall comply with Section C402.5.

Alternatively, where buildings have a vertical fenestration area or skylight area exceeding that allowed in Section C402.4, the building and building thermal envelope shall comply with Section C401.2, Item 1 or Section C401.2, Item 3.

Walk-in coolers, walk-in freezers, refrigerated warehouse coolers and refrigerated warehouse freezers shall comply with Section C403.2.15 or C403.2.16.

C402.1.1 Low-energy buildings. The following low-energy buildings, or portions thereof separated from the remainder of the building by *building thermal envelope* assemblies complying with this section, shall be exempt from the *building thermal envelope* provisions of Section C402.

1. Those with a peak design rate of energy usage less than 3.4 Btu/h $\cdot$ ft^2 (10.7 W/m^2) or 1.0 watt per square foot (10.7 W/m^2) of floor area for space conditioning purposes.

2. Those that do not contain *conditioned space*.

3. Greenhouses.

C402.1.2 Equipment buildings. Buildings that comply with the following shall be exempt from the *building thermal envelope* provisions of this code:

1. Are separate buildings with floor area not more than 500 square feet (50 m^2).

2. Are intended to house electronic equipment with installed equipment power totaling not less than 7 watts per square foot (75 W/m2) and not intended for human occupancy.

3. Have a heating system capacity not greater than (17,000 Btu/hr) (5 kW) and a heating thermostat set point that is restricted to not more than 50°F (10°C).

4. Have an average wall and roof *U-factor* less than 0.200 in *Climate Zones* 1 through 5 and less than 0.120 in *Climate Zones* 6 through 8.

5. Comply with the roof solar reflectance and thermal emittance provisions for *Climate Zone* 1.

C402.1.3 Insulation component *R*-value-based method. *Building thermal envelope* opaque assemblies shall meet the requirements of Sections C402.2 and C402.4 based on the *climate zone* specified in Chapter 3. For opaque portions of the *building thermal envelope* intended to comply on an insulation component *R-value* basis, the *R*-values for insulation in framing cavities, where required, and for continuous insulation, where required, shall be not less than that specified in Table C402.1.3, based on the *climate zone* specified in Chapter 3. Commercial buildings or portions of commercial buildings enclosing Group R occupancies shall use the *R*-values from the "Group R" column of Table C402.1.3. Commercial buildings or portions of commercial buildings enclosing occupancies other than Group R shall use the R-values from the "All other" column of Table C402.1.3. The thermal resistance or *R*-value of the insulating material installed continuously within or on the below-grade exterior walls of the building envelope required in accordance with Table C402.1.3 shall extend to a depth of not less than 10 feet (3048 mm) below the outside finished ground level, or to the level of the lowest floor of the conditioned space enclosed by the below grade wall, whichever is less. Opaque swinging doors shall comply with Table C402.1.4 and opaque nonswinging doors shall comply with Table C402.1.3.

C402.1.4 Assembly *U*-factor, *C*-factor or *F*-factor-based method. Building thermal envelope opaque assemblies intended to comply on an assembly *U*-, *C*- or *F*-factor basis shall have a *U*-, *C*- or *F*-factor not greater than that specified in Table C402.1.4. Commercial buildings or portions of commercial buildings enclosing Group R occupancies shall use the *U*-, *C*- or *F*-factor from the "Group R" column of Table C402.1.4. Commercial buildings or portions of commercial buildings enclosing occupancies other than Group R shall use the *U*-, *C*- or *F*-factor from the "All other" column of Table C402.1.4. The *C*-factor for the below-grade exterior walls of the building envelope, as required in accordance with Table C402.1.4, shall extend to a depth of 10 feet (3048 mm) below the outside finished ground level, or to the level of the lowest floor, whichever is less. Opaque swinging doors shall comply with Table C402.1.4 and opaque non-swinging doors shall comply with Table C402.1.3.

C402.1.4.1 Thermal resistance of cold-formed steel walls. *U*-factors of walls with cold-formed steel studs shall be permitted to be determined in accordance with Equation 4-1:

$U = 1/[R_s + (ER)]$ **(Equation 4-1)**

where:

R_s = The cumulative *R*-value of the wall components along the path of heat transfer, excluding the cavity insulation and steel studs.

ER = The effective *R*-value of the cavity insulation with steel studs.

TABLE C402.1.4.1
EFFECTIVE *R*-VALUES FOR STEEL STUD WALL ASSEMBLIES

NOMINAL STUD DEPTH (inches)	SPACING OF FRAMING (inches)	CAVITY *R*-VALUE (insulation)	CORRECTION FACTOR (F_c)	EFFECTIVE *R*-VALUE (ER) (Cavity *R*-Value × F_c)
3½	16	13	0.46	5.98
		15	0.43	6.45
3½	24	13	0.55	7.15
		15	0.52	7.80
6	16	19	0.37	7.03
		21	0.35	7.35
6	24	19	0.45	8.55
		21	0.43	9.03
8	16	25	0.31	7.75
	24	25	0.38	9.50

C402.1.5 Component performance alternative. Building envelope values and fenestration areas determined in accordance with Equation 4-2 shall be permitted in lieu of compliance with the *U*-, *F*- and *C*-factors in Tables C402.1.3 and C402.1.4 and the maximum allowable fenestration areas in Section C402.4.1.

A + B + C + D + E ≤ Zero **(Equation 4-2)**

where:

A = Sum of the (UA Dif) values for each distinct assembly type of the building thermal envelope, other than slabs on grade and below-grade walls.

    UA Dif = UA Proposed - UA Table.
    UA Proposed = Proposed *U*-value • Area.
    UA Table = (*U*-factor from Table C402.1.3 or Table C402.1.4) • Area.

B = Sum of the (FL Dif) values for each distinct slab-on-grade perimeter condition of the building thermal envelope.

    FL Dif = FL Proposed - FL Table.
    FL Proposed = Proposed *F*-value • Perimeter length.
    FL Table = (*F*-factor specified in Table C402.1.4) • Perimeter length.

C = Sum of the (CA Dif) values for each distinct below-grade wall assembly type of the building thermal envelope.

    CA Dif = CA Proposed - CA Table.
    CA Proposed = Proposed *C*-value • Area.
    CA Table = (Maximum allowable *C*-factor specified in Table C402.1.4) • Area.

Where the proposed vertical glazing area is less than or equal to the maximum vertical glazing area allowed by Section C402.4.1, the value of D (Excess Vertical Glazing Value) shall be zero. Otherwise:

D = (DA • UV) - (DA • U Wall), but not less than zero.

    DA = (Proposed Vertical Glazing Area) - (Vertical Glazing Area allowed by Section C402.4.1).
    UA Wall = Sum of the (UA Proposed) values for each opaque assembly of the exterior wall.
    U Wall = Area-weighted average *U*-value of all above-grade wall assemblies.
    UAV = Sum of the (UA Proposed) values for each vertical glazing assembly.
    UV = UAV/total vertical glazing area.

Where the proposed skylight area is less than or equal to the skylight area allowed by Section C402.4.1, the value of E (Excess Skylight Value) shall be zero. Otherwise:

E = (EA • US) - (EA • U Roof), but not less than zero.

    EA = (Proposed Skylight Area) - (Allowable Skylight Area as specified in Section C402.4.1).
    U Roof = Area-weighted average *U*-value of all roof assemblies.
    UAS = Sum of the (UA Proposed) values for each skylight assembly.
    US = UAS/total skylight area.

COMMERCIAL ENERGY EFFICIENCY

TABLE C402.1.3
OPAQUE THERMAL ENVELOPE INSULATION COMPONENT MINIMUM REQUIREMENTS, R-VALUE METHOD[a]

CLIMATE ZONE	1		2		3		4 EXCEPT MARINE		5 AND MARINE 4		6		7		8	
	All other	Group R	All other	Group R	All other	Group R	All other	Group R	All other	Group R	All other	Group R	All other	Group R	All other	Group R
Roofs																
Insulation entirely above roof deck	R-20ci	R-25ci	R-25ci	R-25ci	R-25ci	R-25ci	R-30ci	R-30ci	R-30ci	R-30ci	R-30ci	R-30ci	R-35ci	R-35ci	R-35ci	R-35ci
Metal buildings[b]	R-19 + R-11 LS	R-19 + R-11 LS	R-19 + R11 LS	R-19 + R-11 LS	R-19 + R-11 LS	R-19 + R-11 LS	R-19 + R-11 LS	R-19 + R-11 LS	R-19 + R-11 LS	R-19 + R-11 LS	R-25 + R-11 LS	R-25 + R-11 LS	R-30 + R-11 LS	R-30 + R-11 LS	R-30 + R-11 LS	R-30 + R-11 LS
Attic and other	R-38	R-38	R-38	R-38	R-38	R-38	R-38	R-38	R-38	R-49	R-49	R-49	R-49	R-49	R-49	R-49
Walls, above grade																
Mass	R-5.7ci[c]	R-5.7ci[c]	R-5.7ci[c]	R-5.7ci[c]	R-7.6ci	R-7.6ci	R-9.5ci	R-11.4ci	R-11.4ci	R-13.3ci	R-13.3ci	R-15.2ci	R-15.2ci	R-15.2ci	R-25ci	R-25ci
Metal building	R-13+ R-6.5ci	R-13 + R-6.5ci	R-13 + R-6.5ci	R-13 + R-6.5ci	R-13 + R-6.5ci	R-13 + R-13ci	R-13 + R-13ci	R-13 + R-13ci	R-13 + R-13ci	R-13 + R-13ci	R-13 + R-13ci	R-13 + R-13ci	R-13 + R-13ci	R-13 + R-19.5ci	R-13 + R-13ci	R-13 + R-19.5ci
Metal framed	R-13 + R-5ci	R-13 + R-5ci	R-13 + R-5ci	R-13 + R-7.5ci	R-13 + R-7.5ci	R-13 + R-7.5ci	R-13 + R-7.5ci	R-13 + R-7.5ci	R-13 + R-7.5ci	R-13 + R-7.5ci	R-13 + R-7.5ci	R-13 + R-7.5ci	R-13 + R-7.5ci	R-13 + R-15.6ci	R-13 + R-7.5ci	R13 + R17.5ci
Wood framed and other	R-13 + R-3.8ci or R-20	R-13 + R-3.8ci or R-20	R-13 + R-3.8ci or R-20	R-13 + R-3.8ci or R-20	R-13 + R-3.8ci or R-20	R-13 + R-3.8ci or R-20	R-13 + R-3.8ci or R-20	R-13 + R-3.8ci or R-20	R-13 + R-3.8ci or R-20	R-13 + R-3.8ci or R-20	R-13 + R-7.5ci or R-20 + R-3.8ci	R-13 + R-7.5ci or R-20 + R-3.8ci	R-13 + R-7.5ci or R-20 + R-3.8ci	R-13 + R-7.5ci or R-20 + R-3.8ci	R13 + R-15.6ci or R-20 + R-10ci	R13 + R-15.6ci or R-20 + R-10ci
Walls, below grade																
Below-grade wall[d]	NR	NR	NR	NR	NR	NR	R-7.5ci	R-7.5ci	R-7.5ci	R-7.5ci	R-7.5ci	R-7.5ci	R-10ci	R-10ci	R-12.5ci	R-12.5ci
Floors																
Mass[e]	NR	NR	R-6.3ci	R-8.3ci	R-10ci	R-10ci	R-10ci	R-10.4ci	R-10ci	R-12.5ci	R-12.5ci	R-12.5ci	R-15ci	R-15ci	R-16.7ci	R-16.7ci
Joist/framing	NR	NR	R-30	R-30	R-30	R-30	R-30	R-30	R-30	R-30	R-30	R-30[f]	R-30[f]	R-30[f]	R-30[f]	R-30[f]
Slab-on-grade floors																
Unheated slabs	NR	NR	NR	NR	NR	NR	R-10 for 24″ below	R-10 for 24″ below	R-10 for 24″ below	R-10 for 24″ below	R-10 for 24″ below	R-10 for 24″ below	R-15 for 24″ below	R-15 for 24″ below	R-20 for 24″ below	R-20 for 24″ below
Heated slabs	R-7.5 for 12″ below	R-7.5 for 12″ below	R-7.5 for 12″ below	R-7.5 for 12″ below	R-10 for 24″ below	R-10 for 24″ below	R-15 for 24″ below	R-15 for 24″ below	R-15 for 36″ below	R-15 for 36″ below	R-15 for 36″ below	R-20 for 48″ below	R-20 for 48″ below	R-20 for 48″ below	R-20 for 48″ below	R-20 for 48″ below
Opaque doors																
Nonswinging	R-4.75	R-4.75	R-4.75	R-4.75	R-4.75	R-4.75	R-4.75	R-4.75	R-4.75	R-4.75	R-4.75	R-4.75	R-4.75	R-4.75	R-4.75	R-4.75

For SI: 1 inch = 25.4 mm, 1 pound per square foot = 4.88 kg/m², 1 pound per cubic foot = 16 kg/m³.

ci = Continuous insulation, NR = No requirement, LS = Liner system.

a. Assembly descriptions can be found in ANSI/ASHRAE/IESNA Appendix A.
b. Where using R-value compliance method, a thermal spacer block shall be provided, otherwise use the U-factor compliance method in Table C402.1.4.
c. R-5.7ci is allowed to be substituted with concrete block walls complying with ASTM C 90, ungrouted or partially grouted at 32 inches or less on center vertically and 48 inches or less on center horizontally, with ungrouted cores filled with materials having a maximum thermal conductivity of 0.44 Btu-in/h-ft²·°F.
d. Where heated slabs are below grade, below-grade walls shall comply with the exterior insulation requirements for heated slabs.
e. "Mass floors" shall include floors weighing not less than:
   1. 35 pounds per square foot of floor surface area; or
   2. 25 pounds per square foot of floor surface area where the material weight is not more than 120 pounds per cubic foot.
f. Steel floor joist systems shall be insulated to R-38.

TABLE C402.1.4
OPAQUE THERMAL ENVELOPE ASSEMBLY MAXIMUM REQUIREMENTS, U-FACTOR METHOD[a, b]

CLIMATE ZONE	1		2		3		4 EXCEPT MARINE		5 AND MARINE 4		6		7		8	
	All other	Group R	All other	Group R	All other	Group R	All other	Group R	All other	Group R	All other	Group R	All other	Group R	All other	Group R
Roofs																
Insulation entirely above roof deck	U-0.048	U-0.039	U-0.039	U-0.039	U-0.039	U-0.039	U-0.032	U-0.032	U-0.032	U-0.032	U-0.032	U-0.032	U-0.028	U-0.028	U-0.028	U-0.028
Metal buildings	U-0.044	U-0.035	U-0.035	U-0.035	U-0.035	U-0.035	U-0.035	U-0.035	U-0.035	U-0.035	U-0.031	U-0.031	U-0.029	U-0.029	U-0.029	U-0.029
Attic and other	U-0.027	U-0.027	U-0.027	U-0.027	U-0.027	U-0.027	U-0.027	U-0.021	U-0.027	U-0.021	U-0.021	U-0.021	U-0.021	U-0.021	U-0.021	U-0.021
Walls, above grade																
Mass	U-0.151	U-0.151	U-0.151	U-0.151	U-0.123	U-0.123	U-0.104	U-0.104	U-0.090	U-0.080	U-0.080	U-0.071	U-0.071	U-0.061	U-0.061	U-0.061
Metal building	U-0.079	U-0.079	U-0.079	U-0.079	U-0.079	U-0.079	U-0.052	U-0.052	U-0.052	U-0.052	U-0.052	U-0.052	U-0.052	U-0.052	U-0.052	U-0.039
Metal framed	U-0.077	U-0.077	U-0.077	U-0.077	U-0.064	U-0.064	U-0.064	U-0.064	U-0.064	U-0.064	U-0.064	U-0.057	U-0.064	U-0.052	U-0.045	U-0.045
Wood framed and other[c]	U-0.064	U-0.064	U-0.064	U-0.064	U-0.064	U-0.064	U-0.064	U-0.064	U-0.064	U-0.064	U-0.051	U-0.051	U-0.051	U-0.051	U-0.036	U-0.036
Walls, below grade																
Below-grade wall[c]	C-1.140[e]	C-1.140[e]	C-1.140[e]	C-1.140[e]	C-1.140[e]	C-1.140[e]	C-0.119	C-0.119	C-0.119	C-0.119	C-0.119	C-0.119	C-0.092	C-0.092	C-0.092	C-0.092
Floors																
Mass[d]	U-0.322[e]	U-0.322[e]	U-0.107	U-0.087	U-0.076	U-0.076	U-0.076	U-0.074	U-0.074	U-0.064	U-0.064	U-0.057	U-0.055	U-0.051	U-0.055	U-0.051
Joist/framing	U-0.066[e]	U-0.066[e]	U-0.033	U-0.033	U-0.033	U-0.033	U-0.033	U-0.033	U-0.033	U-0.033	U-0.033	U-0.033	U-0.033	U-0.033	U-0.033	U-0.033
Slab-on-grade floors																
Unheated slabs	F-0.73[e]	F-0.73[e]	F-0.73[e]	F-0.73[e]	F-0.73[e]	F-0.73[e]	F-0.54	F-0.54	F-0.54	F-0.54	F-0.54	F-0.52	F-0.40	F-0.40	F-0.40	F-0.40
Heated slabs[f]	F-0.70	F-0.70	F-0.70	F-0.70	F-0.70	F-0.70	F-0.65	F-0.65	F-0.65	F-0.65	F-0.58	F-0.58	F-0.55	F-0.55	F-0.55	F-0.55
Opaque doors																
Swinging	U-0.61	U-0.61	U-0.61	U-0.51	U-0.61	U-0.61	U-0.61	U-0.61	U-0.37	U-0.37	U-0.37	U-0.37	U-0.37	U-0.37	U-0.37	U-0.37

For SI: 1 pound per square foot = 4.88 kg/m², 1 pound per cubic foot = 16 kg/m³.

ci = Continuous insulation, NR = No requirement, LS = Liner system.

a. Use of Opaque assembly U-factors, C-factors, and F-factors from ANSI/ASHRAE/IESNA 90.1 Appendix A shall be permitted, provided the construction, excluding the cladding system on walls, complies with the appropriate construction details from ANSI/ASHRAE/IESNA 90.1 Appendix A.
b. Opaque assembly U-factors based on designs tested in accordance with ASTM C1363 shall be permitted. The R-value of continuous insulation shall be permitted to be added to or subtracted from the original tested design.
c. Where heated slabs are below grade, below-grade walls shall comply with the F-factor requirements for heated slabs.
d. "Mass floors" shall include floors weighing not less than:
  1. 35 pounds per square foot of floor surface area; or
  2. 25 pounds per square foot of floor surface area where the material weight is not more than 120 pounds per cubic foot.
e. These C-, F- and U-factors are based on assemblies that are not required to contain insulation.
f. Evidence of compliance with the F-factors indicated in the table for heated slabs shall be demonstrated by the application of the unheated slab F-factors and R-values derived from ASHRAE 90.1 Appendix A.

C402.2 Specific building thermal envelope insulation requirements (Prescriptive). Insulation in building thermal envelope opaque assemblies shall comply with Sections C402.2.1 through C402.2.6 and Table C402.1.3.

C402.2.1 Multiple layers of continuous insulation board. Where two or more layers of continuous insulation board are used in a construction assembly, the continuous insulation boards shall be installed in accordance with Section C303.2. Where the continuous insulation board manufacturer's instructions do not address installation of two or more layers, the edge joints between each layer of continuous insulation boards shall be staggered.

C402.2.2 Roof assembly. The minimum thermal resistance (R-value) of the insulating material installed either between the roof framing or continuously on the roof assembly shall be as specified in Table C402.1.3, based on construction materials used in the roof assembly. Skylight curbs shall be insulated to the level of roofs with insulation entirely above deck or R-5, whichever is less.

Exceptions:

1. Continuously insulated roof assemblies where the thickness of insulation varies 1 inch (25 mm) or less and where the area-weighted U-factor is equivalent to the same assembly with the R-value specified in Table C402.1.3.
2. Where tapered insulation is used with insulation entirely above deck, the R-value where the insulation thickness varies 1 inch (25 mm) or less from the minimum thickness of tapered insulation shall comply with the R-value specified in Table C402.1.3.
3. Unit skylight curbs included as a component of a skylight listed and labeled in accordance with NFRC 100 shall not be required to be insulated.

Insulation installed on a suspended ceiling with removable ceiling tiles shall not be considered part of the minimum thermal resistance of the roof insulation.

C402.2.3 Thermal resistance of above-grade walls. The minimum thermal resistance (R-value) of materials installed in the wall cavity between framing members and continuously on the walls shall be as specified in Table C402.1.3, based on framing type and construction materials used in the wall assembly. The R-value of integral insulation installed in concrete masonry units shall not be used in determining compliance with Table C402.1.3.

"Mass walls" shall include walls:

1. Weighing not less than 35 psf (170 kg/m^2) of wall surface area.
2. Weighing not less than 25 psf (120 kg/m^2) of wall surface area where the material weight is not more than 120 pcf (1900 kg/m^3).
3. Having a heat capacity exceeding 7 Btu/ft$^2 \cdot$ °F (144 kJ/m$^2 \cdot$ K).
4. Having a heat capacity exceeding 5 Btu/ft$^2 \cdot$ °F (103 kJ/m$^2 \cdot$ K), where the material weight is not more than 120 pcf (1900 kg/m^3).

C402.2.4 Floors. The thermal properties (component R-values or assembly U-, C- or F-factors) of floor assemblies over outdoor air or unconditioned space shall be as specified in Table C402.1.3 or C402.1.4 based on the construction materials used in the floor assembly. Floor framing cavity insulation or structural slab insulation shall be installed to maintain permanent contact with the underside of the subfloor decking or structural slabs.

Exceptions:

1. The floor framing cavity insulation or structural slab insulation shall be permitted to be in contact with the top side of sheathing or continuous insulation installed on the bottom side of floor assemblies where combined with insulation that meets or exceeds the minimum R-value in Table C402.1.3 for "Metal framed" or "Wood framed and other" values for "Walls, Above Grade" and extends from the bottom to the top of all perimeter floor framing or floor assembly members.
2. Insulation applied to the underside of concrete floor slabs shall be permitted an airspace of not more than 1 inch (25 mm) where it turns up and is in contact with the underside of the floor under walls associated with the *building thermal envelope*.

C402.2.5 Slabs-on-grade perimeter insulation. Where the slab on grade is in contact with the ground, the minimum thermal resistance (R-value) of the insulation around the perimeter of unheated or heated slab-on-grade floors designed in accordance with the R-value method of Section C402.1.3 shall be as specified in Table C402.1.3. The insulation shall be placed on the outside of the foundation or on the inside of the foundation wall. The insulation shall extend downward from the top of the slab for a minimum distance as shown in the table or to the top of the footing, whichever is less, or downward to at least the bottom of the slab and then horizontally to the interior or exterior for the total distance shown in the table. Insulation extending away from the building shall be protected by pavement or by not less than of 10 inches (254 mm) of soil.

Exception: Where the slab-on-grade floor is greater than 24 inches (61 mm) below the finished exterior grade, perimeter insulation is not required.

C402.2.6 Insulation of radiant heating systems. *Radiant heating system* panels, and their associated components that are installed in interior or exterior assemblies shall be insulated with a minimum of R-3.5 (0.62 m^2/K $\cdot$ W) on all surfaces not facing the space being heated. *Radiant heating system* panels that are installed in the *building thermal envelope* shall be separated from the exterior of the building or unconditioned or exempt spaces by not less than the R-value of insulation installed in the opaque assembly in which they are installed or the assembly shall comply with Section C402.1.4.

Exception: Heated slabs on grade insulated in accordance with Section C402.2.5.

C402.3 Roof solar reflectance and thermal emittance. Low-sloped roofs directly above cooled conditioned spaces in *Climate Zones* 1, 2 and 3 shall comply with one or more of the options in Table C402.3.

Exceptions: The following roofs and portions of roofs are exempt from the requirements of Table C402.3:

1. Portions of the roof that include or are covered by the following:
    1.1. Photovoltaic systems or components.
    1.2. Solar air or water-heating systems or components.
    1.3. Roof gardens or landscaped roofs.
    1.4. Above-roof decks or walkways.
    1.5. Skylights.
    1.6. HVAC systems and components, and other opaque objects mounted above the roof.
2. Portions of the roof shaded during the peak sun angle on the summer solstice by permanent features of the building or by permanent features of adjacent buildings.
3. Portions of roofs that are ballasted with a minimum stone ballast of 17 pounds per square foot [74 kg/m^2] or 23 psf [117 kg/m^2] pavers.
4. Roofs where not less than 75 percent of the roof area complies with one or more of the exceptions to this section.

C402.3.1 Aged roof solar reflectance. Where an aged solar reflectance required by Section C402.3 is not available, it shall be determined in accordance with Equation 4-3.

$$R_{aged} = [0.2 + 0.7(R_{initial} - 0.2)] \quad \text{(Equation 4-3)}$$

where:

R_{aged} = The aged solar reflectance.

$R_{initial}$ = The initial solar reflectance determined in accordance with CRRC-1 Standard.

C402.4 Fenestration (Prescriptive). Fenestration shall comply with Sections C402.4 through C402.4.4 and Table C402.4. Daylight responsive controls shall comply with this section and Section C405.2.3.1.

C402.4.1 Maximum area. The vertical fenestration area (not including opaque doors and opaque spandrel panels) shall not be greater than 30 percent of the gross above-grade wall area. The skylight area shall not be greater than 3 percent of the gross roof area.

TABLE C402.3
MINIMUM ROOF REFLECTANCE AND EMITTANCE OPTIONS[a]

Three-year aged solar reflectance[b] of 0.55 and 3-year aged thermal emittance[c] of 0.75
Three-year-aged solar reflectance index[d] of 64

a. The use of area-weighted averages to comply with these requirements shall be permitted. Materials lacking 3-year-aged tested values for either solar reflectance or thermal emittance shall be assigned both a 3-year-aged solar reflectance in accordance with Section C402.3.1 and a 3-year-aged thermal emittance of 0.90.
b. Aged solar reflectance tested in accordance with ASTM C 1549, ASTM E 903 or ASTM E 1918 or CRRC-1 Standard.
c. Aged thermal emittance tested in accordance with ASTM C 1371 or ASTM E 408 or CRRC-1 Standard.
d. Solar reflectance index (SRI) shall be determined in accordance with ASTM E 1980 using a convection coefficient of 2.1 Btu/h · ft^2 · °F (12W/m^2 · K). Calculation of aged SRI shall be based on aged tested values of solar reflectance and thermal emittance.

TABLE C402.4
BUILDING ENVELOPE FENESTRATION MAXIMUM *U*-FACTOR AND SHGC REQUIREMENTS

CLIMATE ZONE	1		2		3		4 EXCEPT MARINE		5 AND MARINE 4		6		7		8	
Vertical fenestration																
***U*-factor**																
Fixed fenestration	0.50		0.50		0.46		0.38		0.38		0.36		0.29		0.29	
Operable fenestration	0.65		0.65		0.60		0.45		0.45		0.43		0.37		0.37	
Entrance doors	1.10		0.83		0.77		0.77		0.77		0.77		0.77		0.77	
SHGC																
Orientation[a]	SEW	N	SEW	N	SEW	N	SEW	N	SEW	N	SEW	N	SEW	N	SEW	N
PF < 0.2	0.25	0.33	0.25	0.33	0.25	0.33	0.40	0.53	0.40	0.53	0.40	0.53	0.45	NR	0.45	N
0.2 ≤ PF < 0.5	0.30	0.37	0.30	0.37	0.30	0.37	0.48	0.58	0.48	0.58	0.48	0.58	NR	NR	NR	NR
PF ≥ 0.5	0.40	0.40	0.40	0.40	0.40	0.40	0.64	0.64	0.64	0.64	0.64	0.64	NR	NR	NR	NR
Skylights																
U-factor	0.75		0.65		0.55		0.50		0.50		0.50		0.50		0.50	
SHGC	0.35		0.35		0.35		0.40		0.40		0.40		NR		NR	

NR = No requirement, PF = Projection factor.
a. "N" indicates vertical fenestration oriented within 45 degrees of true north. "SEW" indicates orientations other than "N." For buildings in the southern hemisphere, reverse south and north. Buildings located at less than 23.5 degrees latitude shall use SEW for all orientations.

C402.4.1.1 Increased vertical fenestration area with daylight responsive controls. In *Climate Zones* 1 through 6, not more than 40 percent of the gross above-grade wall area shall be permitted to be vertical fenestration, provided all of the following requirements are met:

1. In buildings not greater than two stories above grade, not less than 50 percent of the net floor area is within a *daylight zone*.
2. In buildings three or more stories above grade, not less than 25 percent of the net floor area is within a *daylight zone*.
3. *Daylight responsive controls* complying with Section C405.2.3.1 are installed in *daylight zones*.
4. Visible transmittance (VT) of vertical fenestration is not less than 1.1 times solar heat gain coefficient (SHGC).

   Exception: Fenestration that is outside the scope of NFRC 200 is not required to comply with Item 4.

C402.4.1.2 Increased skylight area with daylight responsive controls. The skylight area shall be permitted to be not more than 5 percent of the roof area provided *daylight responsive controls* complying with Section C405.2.3.1 are installed in *daylight zones* under skylights.

C402.4.2 Minimum skylight fenestration area. In an enclosed space greater than 2,500 square feet (232 m^2) in floor area, directly under a roof with not less than 75 percent of the ceiling area with a ceiling height greater than 15 feet (4572 mm), and used as an office, lobby, atrium, concourse, corridor, storage space, gymnasium/exercise center, convention center, automotive service area, space where manufacturing occurs, nonrefrigerated warehouse, retail store, distribution/sorting area, transportation depot or workshop, the total *daylight zone* under skylights shall be not less than half the floor area and shall provide one of the following:

1. A minimum skylight area to *daylight zone* under skylights of not less than 3 percent where all skylights have a VT of at least 0.40 as determined in accordance with Section C303.1.3.
2. A minimum skylight effective aperture of at least 1 percent, determined in accordance with Equation 4-4.

$$\text{Skylight Effective Aperture} = \frac{0.85 \cdot \text{Skylight Area} \cdot \text{Skylight VT} \cdot \text{WF}}{\text{Daylight zone under skylight}}$$

(Equation 4-4)

where:

Skylight area	= Total fenestration area of skylights.
Skylight VT	= Area weighted average visible transmittance of skylights.
WF	= Area weighted average well factor, where well factor is 0.9 if light well depth is less than 2 feet (610 mm), or 0.7 if light well depth is 2 feet (610 mm) or greater.
Light well depth	= Measure vertically from the underside of the lowest point of the skylight glazing to the ceiling plane under the skylight.

Exception: Skylights above *daylight zones* of enclosed spaces are not required in:

1. Buildings in *Climate Zones* 6 through 8.
2. Spaces where the designed *general lighting* power densities are less than 0.5 W/ft^2 (5.4 W/m^2).
3. Areas where it is documented that existing structures or natural objects block direct beam sunlight on at least half of the roof over the enclosed area for more than 1,500 daytime hours per year between 8 a.m. and 4 p.m.
4. Spaces where the *daylight zone* under rooftop monitors is greater than 50 percent of the enclosed space floor area.
5. Spaces where the total area minus the area of *daylight zones* adjacent to vertical fenestration is less than 2,500 square feet (232 m^2), and where the lighting is controlled according to Section C405.2.3.

C402.4.2.1 Lighting controls in daylight zones under skylights. *Daylight responsive controls* complying with Section C405.2.3.1 shall be provided to control all electric lights within *daylight zones* under skylights.

C402.4.2.2 Haze factor. Skylights in office, storage, automotive service, manufacturing, nonrefrigerated warehouse, retail store and distribution/sorting area spaces shall have a glazing material or diffuser with a haze factor greater than 90 percent when tested in accordance with ASTM D 1003.

Exception: Skylights designed and installed to exclude direct sunlight entering the occupied space by the use of fixed or automated baffles or the geometry of skylight and light well.

C402.4.3 Maximum *U*-factor and SHGC. The maximum *U*-factor and solar heat gain coefficient (SHGC) for fenestration shall be as specified in Table C402.4.

The window projection factor shall be determined in accordance with Equation 4-5.

$PF = A/B$ **(Equation 4-5)**

where:

PF = Projection factor (decimal).

A = Distance measured horizontally from the furthest continuous extremity of any overhang, eave or permanently attached shading device to the vertical surface of the glazing.

B = Distance measured vertically from the bottom of the glazing to the underside of the overhang, eave or permanently attached shading device.

Where different windows or glass doors have different PF values, they shall each be evaluated separately.

C402.4.3.1 Increased skylight SHGC. In *Climate Zones* 1 through 6, skylights shall be permitted a maximum SHGC of 0.60 where located above *daylight zones* provided with *daylight responsive controls*.

C402.4.3.2 Increased skylight U-factor. Where skylights are installed above *daylight zones* provided with *daylight responsive controls*, a maximum *U*-factor of 0.9 shall be permitted in *Climate Zones* 1 through 3 and a maximum *U*-factor of 0.75 shall be permitted in *Climate Zones* 4 through 8.

C402.4.3.3 Dynamic glazing. Where *dynamic glazing* is intended to satisfy the SHGC and VT requirements of Table C402.4, the ratio of the higher to lower labeled SHGC shall be greater than or equal to 2.4, and the *dynamic glazing* shall be automatically controlled to modulate the amount of solar gain into the space in multiple steps. Dynamic glazing shall be considered separately from other fenestration, and area-weighted averaging with other fenestration that is not dynamic glazing shall not be permitted.

> **Exception:** Dynamic glazing is not required to comply with this section where both the lower and higher labeled SHGC already comply with the requirements of Table C402.4.

C402.4.3.4 Area-weighted U-factor. An area-weighted average shall be permitted to satisfy the *U*-factor requirements for each fenestration product category listed in Table C402.4. Individual fenestration products from different fenestration product categories listed in Table C402.4 shall not be combined in calculating area-weighted average *U*-factor.

C402.4.4 Doors. *Opaque doors* shall comply with the applicable requirements for doors as specified in Tables C402.1.3 and C402.1.4 and be considered part of the gross area of above-grade walls that are part of the building *thermal envelope*. Other doors shall comply with the provisions of Section C402.4.3 for vertical fenestration.

C402.5 Air leakage—thermal envelope (Mandatory). The *thermal envelope* of buildings shall comply with Sections C402.5.1 through C402.5.8, or the building *thermal envelope* shall be tested in accordance with ASTM E 779 at a pressure differential of 0.3 inch water gauge (75 Pa) or an equivalent method approved by the code official and deemed to comply with the provisions of this section when the tested air leakage rate of the building thermal envelope is not greater than 0.40 cfm/ft^2 (0.2 L/s · m^2). Where compliance is based on such testing, the building shall also comply with Sections C402.5.5, C402.5.6 and C402.5.7.

C402.5.1 Air barriers. A continuous air barrier shall be provided throughout the building thermal envelope. The air barriers shall be permitted to be located on the inside or outside of the building envelope, located within the assemblies composing the envelope, or any combination thereof. The air barrier shall comply with Sections C402.5.1.1 and C402.5.1.2.

> **Exception:** Air barriers are not required in buildings located in *Climate Zone* 2B.

C402.5.1.1 Air barrier construction. The *continuous air barrier* shall be constructed to comply with the following:

1. The air barrier shall be continuous for all assemblies that are the thermal envelope of the building and across the joints and assemblies.

2. Air barrier joints and seams shall be sealed, including sealing transitions in places and changes in materials. The joints and seals shall be securely installed in or on the joint for its entire length so as not to dislodge, loosen or otherwise impair its ability to resist positive and negative pressure from wind, stack effect and mechanical ventilation.

3. Penetrations of the air barrier shall be caulked, gasketed or otherwise sealed in a manner compatible with the construction materials and location. Joints and seals associated with penetrations shall be sealed in the same manner or taped or covered with moisture vapor-permeable wrapping material. Sealing materials shall be appropriate to the construction materials being sealed and shall be securely installed around the penetration so as not to dislodge, loosen or otherwise impair the penetrations' ability to resist positive and negative pressure from wind, stack effect and mechanical ventilation. Sealing of concealed fire sprinklers, where required, shall be in a manner that is recommended by the manufacturer. Caulking or other adhesive sealants shall not be used to fill voids between fire sprinkler cover plates and walls or ceilings.

4. Recessed lighting fixtures shall comply with Section C402.5.8. Where similar objects are installed that penetrate the air barrier, provisions shall be made to maintain the integrity of the air barrier.

C402.5.1.2 Air barrier compliance options. A continuous air barrier for the opaque building envelope shall comply with Section C402.5.1.2.1 or C402.5.1.2.2.

C402.5.1.2.1 Materials. Materials with an air permeability not greater than 0.004 cfm/ft^2 (0.02 L/s · m^2) under a pressure differential of 0.3 inch water gauge (75 Pa) when tested in accordance with ASTM E 2178 shall comply with this section. Materials in Items 1 through 16 shall be deemed to comply with this section, provided joints are sealed and materials are installed as air barriers in accordance with the manufacturer's instructions.

1. Plywood with a thickness of not less than $^3/_8$ inch (10 mm).

2. Oriented strand board having a thickness of not less than $^3/_8$ inch (10 mm).

3. Extruded polystyrene insulation board having a thickness of not less than $^1/_2$ inch (12.7 mm).

4. Foil-back polyisocyanurate insulation board having a thickness of not less than $^1/_2$ inch (12.7 mm).
5. Closed-cell spray foam a minimum density of 1.5 pcf (2.4 kg/m^3) having a thickness of not less than $1^1/_2$ inches (38 mm).
6. Open-cell spray foam with a density between 0.4 and 1.5 pcf (0.6 and 2.4 kg/m^3) and having a thickness of not less than 4.5 inches (113 mm).
7. Exterior or interior gypsum board having a thickness of not less than $^1/_2$ inch (12.7 mm).
8. Cement board having a thickness of not less than $^1/_2$ inch (12.7 mm).
9. Built-up roofing membrane.
10. Modified bituminous roof membrane.
11. Fully adhered single-ply roof membrane.
12. A Portland cement/sand parge, or gypsum plaster having a thickness of not less than $^5/_8$ inch (15.9 mm).
13. Cast-in-place and precast concrete.
14. Fully grouted concrete block masonry.
15. Sheet steel or aluminum.
16. Solid or hollow masonry constructed of clay or shale masonry units.

C402.5.1.2.2 Assemblies. Assemblies of materials and components with an average air leakage not greater than 0.04 cfm/ft^2 (0.2 L/s · m^2) under a pressure differential of 0.3 inch of water gauge (w.g.)(75 Pa) when tested in accordance with ASTM E 2357, ASTM E 1677 or ASTM E 283 shall comply with this section. Assemblies listed in Items 1 through 3 shall be deemed to comply, provided joints are sealed and the requirements of Section C402.5.1.1 are met.

1. Concrete masonry walls coated with either one application of block filler or two applications of a paint or sealer coating.
2. Masonry walls constructed of clay or shale masonry units with a nominal width of 4 inches (102 mm) or more.
3. A Portland cement/sand parge, stucco or plaster not less than $^1/_2$ inch (12.7 mm) in thickness.

C402.5.2 Air leakage of fenestration. The air leakage of fenestration assemblies shall meet the provisions of Table C402.5.2. Testing shall be in accordance with the applicable reference test standard in Table C402.5.2 by an accredited, independent testing laboratory and *labeled* by the manufacturer.

Exceptions:

1. Field-fabricated fenestration assemblies that are sealed in accordance with Section C402.5.1.
2. Fenestration in buildings that comply with the testing alternative of Section C402.5 are not required to meet the air leakage requirements in Table C402.5.2.

C402.5.3 Rooms containing fuel-burning appliances. In *Climate Zones* 3 through 8, where open combustion air ducts provide combustion air to open combustion space conditioning fuel-burning appliances, the appliances and combustion air openings shall be located outside of the *building thermal envelope* or enclosed in a room isolated from inside the thermal envelope. Such rooms shall be sealed and insulated in accordance with the envelope requirements of Table C402.1.3 or C402.1.4, where the walls, floors and ceilings shall meet the minimum of the below-grade wall *R*-value requirement. The door into the room shall be fully gasketed, and any water lines and ducts in the room insulated in accordance with Section C403. The combustion air duct shall be insulated, where it passes through conditioned space, to a minimum of R-8.

Exceptions:

1. Direct vent appliances with both intake and exhaust pipes installed continuous to the outside.
2. Fireplaces and stoves complying with Sections 901 through 905 of the *International Mechanical Code*, and Section 2111.13 of the *International Building Code*.

C402.5.4 Doors and access openings to shafts, chutes, stairways and elevator lobbies. Doors and access openings from conditioned space to shafts, chutes stairways and elevator lobbies not within the scope of the fenestration assemblies covered by Section C402.5.2 shall be gasketed, weatherstripped or sealed.

Exceptions:

1. Door openings required to comply with Section 716 or 716.5 of the *International Building Code*.
2. Doors and door openings required to comply with UL 1784 by the *International Building Code*.

TABLE C402.5.2
MAXIMUM AIR LEAKAGE RATE FOR FENESTRATION ASSEMBLIES

FENESTRATION ASSEMBLY	MAXIMUM RATE (CFM/FT^2)	TEST PROCEDURE
Windows	0.20^a	AAMA/WDMA/CSA101/I.S.2/A440 or NFRC 400
Sliding doors	0.20^a	
Swinging doors	0.20^a	
Skylights – with condensation weepage openings	0.30	
Skylights – all other	0.20^a	
Curtain walls	0.06	NFRC 400 or ASTM E 283 at 1.57 psf (75 Pa)
Storefront glazing	0.06	
Commercial glazed swinging entrance doors	1.00	
Revolving doors	1.00	
Garage doors	0.40	ANSI/DASMA 105, NFRC 400, or ASTM E 283 at 1.57 psf (75 Pa)
Rolling doors	1.00	
High-speed doors	1.30	

For SI: 1 cubic foot per minute = 0.47 L/s, 1 square foot = 0.093 m^2.

a. The maximum rate for windows, sliding and swinging doors, and skylights is permitted to be 0.3 cfm per square foot of fenestration or door area when tested in accordance with AAMA/WDMA/CSA101/I.S.2/A440 at 6.24 psf (300 Pa).

C402.5.5 Air intakes, exhaust openings, stairways and shafts. Stairway enclosures, elevator shaft vents and other outdoor air intakes and exhaust openings integral to the building envelope shall be provided with dampers in accordance with Section C403.2.4.3.

C402.5.6 Loading dock weatherseals. Cargo doors and loading dock doors shall be equipped with weatherseals to restrict infiltration when vehicles are parked in the doorway.

C402.5.7 Vestibules. Building entrances shall be protected with an enclosed vestibule, with all doors opening into and out of the vestibule equipped with self-closing devices. Vestibules shall be designed so that in passing through the vestibule it is not necessary for the interior and exterior doors to open at the same time. The installation of one or more revolving doors in the building entrance shall not eliminate the requirement that a vestibule be provided on any doors adjacent to revolving doors.

> **Exceptions:** Vestibules are not required for the following:
>
> 1. Buildings in *Climate Zones* 1 and 2.
>
> 2. Doors not intended to be used by the public, such as doors to mechanical or electrical equipment rooms, or intended solely for employee use.
>
> 3. Doors opening directly from a *sleeping unit* or dwelling unit.
>
> 4. Doors that open directly from a space less than 3,000 square feet (298 m^2) in area.
>
> 5. Revolving doors.
>
> 6. Doors used primarily to facilitate vehicular movement or material handling and adjacent personnel doors.
>
> 7. Doors that have an air curtain with a velocity of not less than 6.56 feet per second (2 m/s) at the floor that have been tested in accordance with ANSI/AMCA 220 and installed in accordance with the manufacturer's instructions. Manual or automatic controls shall be provided that will operate the air curtain with the opening and closing of the door. Air curtains and their controls shall comply with Section C408.2.3.

C402.5.8 Recessed lighting. Recessed luminaires installed in the *building thermal envelope* shall be all of the following:

1. IC-rated.

2. Labeled as having an air leakage rate of not more 2.0 cfm (0.944 L/s) when tested in accordance with ASTM E 283 at a 1.57 psf (75 Pa) pressure differential.

3. Sealed with a gasket or caulk between the housing and interior wall or ceiling covering.

SECTION C403
BUILDING MECHANICAL SYSTEMS

C403.1 General. Mechanical systems and equipment serving the building heating, cooling or ventilating needs shall comply with Section C403.2 and shall comply with Sections C403.3 and C403.4 based on the equipment and systems provided.

Walk-in coolers, walk-in freezers, refrigerated warehouse coolers and refrigerated warehouse freezers shall comply with Section C403.2.15 or C403.2.16.

C403.2 Provisions applicable to all mechanical systems (Mandatory). Mechanical systems and equipment serving the building heating, cooling or ventilating needs shall comply with Sections C403.2.1 through C403.2.16.

C403.2.1 Calculation of heating and cooling loads. Design loads associated with heating, ventilating and air conditioning of the building shall be determined in accordance with ANSI/ASHRAE/ACCA Standard 183 or by an *approved* equivalent computational procedure using the design parameters specified in Chapter 3. Heating and cooling loads shall be adjusted to account for load reductions that are achieved where energy recovery systems are utilized in the HVAC system in accordance with the ASHRAE *HVAC Systems and Equipment Handbook* by an approved equivalent computational procedure.

C403.2.2 Equipment sizing. The output capacity of heating and cooling equipment shall be not greater than the loads calculated in accordance with Section C403.2.1. A single piece of equipment providing both heating and cooling shall satisfy this provision for one function with the capacity for the other function as small as possible, within available equipment options.

> **Exceptions:**
>
> 1. Required standby equipment and systems provided with controls and devices that allow such systems or equipment to operate automatically only when the primary equipment is not operating.
>
> 2. Multiple units of the same equipment type with combined capacities exceeding the design load and provided with controls that have the capability to sequence the operation of each unit based on load.

C403.2.3 HVAC equipment performance requirements. Equipment shall meet the minimum efficiency requirements of Tables C403.2.3(1), C403.2.3(2), C403.2.3(3), C403.2.3(4), C403.2.3(5), C403.2.3(6), C403.2.3(7), C403.2.3(8) and C403.2.3(9) when tested and rated in accordance with the applicable test procedure. Plate-type liquid-to-liquid heat exchangers shall meet the minimum requirements of Table C403.2.3(10). The efficiency shall be verified through certification under an *approved* certification program or, where a certification program does not exist, the equipment efficiency ratings shall be supported by data furnished by the manufacturer. Where multiple rating conditions or performance requirements are provided, the equipment shall satisfy all stated requirements. Where components, such as indoor or outdoor coils, from different manufacturers are used, calculations and supporting data shall be furnished by the designer that demonstrates that the combined efficiency of the specified components meets the requirements herein.

TABLE C403.2.3(1)
MINIMUM EFFICIENCY REQUIREMENTS:
ELECTRICALLY OPERATED UNITARY AIR CONDITIONERS AND CONDENSING UNITS

EQUIPMENT TYPE	SIZE CATEGORY	HEATING SECTION TYPE	SUBCATEGORY OR RATING CONDITION	MINIMUM EFFICIENCY		TEST PROCEDURE[a]
				Before 1/1/2016	As of 1/1/2016	
Air conditioners, air cooled	< 65,000 Btu/h[b]	All	Split System	13.0 SEER	13.0 SEER	AHRI 210/240
			Single Package	13.0 SEER	14.0 SEERc	
Through-the-wall (air cooled)	≤ 30,000 Btu/h[b]	All	Split system	12.0 SEER	12.0 SEER	
			Single Package	12.0 SEER	12.0 SEER	
Small-duct high-velocity (air cooled)	< 65,000 Btu/h[b]	All	Split System	11.0 SEER	11.0 SEER	
Air conditioners, air cooled	≥ 65,000 Btu/h and < 135,000 Btu/h	Electric Resistance (or None)	Split System and Single Package	11.2 EER 11.4 IEER	11.2 EER 12.8 IEER	AHRI 340/360
		All other	Split System and Single Package	11.0 EER 11.2 IEER	11.0 EER 12.6 IEER	
	≥ 135,000 Btu/h and < 240,000 Btu/h	Electric Resistance (or None)	Split System and Single Package	11.0 EER 11.2 IEER	11.0 EER 12.4 IEER	
		All other	Split System and Single Package	10.8 EER 11.0 IEER	10.8 EER 12.2 IEER	
	≥ 240,000 Btu/h and < 760,000 Btu/h	Electric Resistance (or None)	Split System and Single Package	10.0 EER 10.1 IEER	10.0 EER 11.6 IEER	
		All other	Split System and Single Package	9.8 EER 9.9 IEER	9.8 EER 11.4 IEER	
	≥ 760,000 Btu/h	Electric Resistance (or None)	Split System and Single Package	9.7 EER 9.8 IEER	9.7 EER 11.2 IEER	
		All other	Split System and Single Package	9.5 EER 9.6 IEER	9.5 EER 11.0 IEER	
Air conditioners, water cooled	< 65,000 Btu/h[b]	All	Split System and Single Package	12.1 EER 12.3 IEER	12.1 EER 12.3 IEER	AHRI 210/240
	≥ 65,000 Btu/h and < 135,000 Btu/h	Electric Resistance (or None)	Split System and Single Package	12.1 EER 12.3 IEER	12.1 EER 13.9 IEER	AHRI 340/360
		All other	Split System and Single Package	11.9 EER 12.1 IEER	11.9 EER 13.7 IEER	
	≥ 135,000 Btu/h and < 240,000 Btu/h	Electric Resistance (or None)	Split System and Single Package	12.5 EER 12.5 IEER	12.5 EER 13.9 IEER	
		All other	Split System and Single Package	12.3 EER 12.5 IEER	12.3 EER 13.7 IEER	
	≥ 240,000 Btu/h and < 760,000 Btu/h	Electric Resistance (or None)	Split System and Single Package	12.4 EER 12.6 IEER	12.4 EER 13.6 IEER	
		All other	Split System and Single Package	12.2 EER 12.4 IEER	12.2 EER 13.4 IEER	
	≥ 760,000 Btu/h	Electric Resistance (or None)	Split System and Single Package	12.2 EER 12.4 IEER	12.2 EER 13.5 IEER	
		All other	Split System and Single Package	12.0 EER 12.2 IEER	12.0 EER 13.3 IEER	

(continued)

TABLE C403.2.3(1)—continued
MINIMUM EFFICIENCY REQUIREMENTS:
ELECTRICALLY OPERATED UNITARY AIR CONDITIONERS AND CONDENSING UNITS

EQUIPMENT TYPE	SIZE CATEGORY	HEATING SECTION TYPE	SUB-CATEGORY OR RATING CONDITION	MINIMUM EFFICIENCY Before 1/1/2016	MINIMUM EFFICIENCY As of 1/1/2016	TEST PROCEDURE[a]
Air conditioners, evaporatively cooled	< 65,000 Btu/h[b]	All	Split System and Single Package	12.1 EER 12.3 IEER	12.1 EER 12.3 IEER	AHRI 210/240
	≥ 65,000 Btu/h and < 135,000 Btu/h	Electric Resistance (or None)	Split System and Single Package	12.1 EER 12.3 IEER	12.1 EER 12.3 IEER	AHRI 340/360
		All other	Split System and Single Package	11.9 EER 12.1 IEER	11.9 EER 12.1 IEER	
	≥ 135,000 Btu/h and < 240,000 Btu/h	Electric Resistance (or None)	Split System and Single Package	12.0 EER 12.2 IEER	12.0 EER 12.2 IEER	
		All other	Split System and Single Package	11.8 EER 12.0 IEER	11.8 EER 12.0 IEER	
	≥ 240,000 Btu/h and < 760,000 Btu/h	Electric Resistance (or None)	Split System and Single Package	11.9 EER 12.1 IEER	11.9 EER 12.1 IEER	
		All other	Split System and Single Package	11.7 EER 11.9 IEER	11.7 EER 11.9 IEER	
	≥ 760,000 Btu/h	Electric Resistance (or None)	Split System and Single Package	11.7 EER 11.9 IEER	11.7 EER 11.9 IEER	
		All other	Split System and Single Package	11.5 EER 11.7 IEER	11.5 EER 11.7 IEER	
Condensing units, air cooled	≥ 135,000 Btu/h			10.5 EER 11.8 IEER	10.5 EER 11.8 IEER	AHRI 365
Condensing units, water cooled	≥ 135,000 Btu/h			13.5 EER 14.0 IEER	13.5 EER 14.0 IEER	
Condensing units, evaporatively cooled	≥ 135,000 Btu/h			13.5 EER 14.0 IEER	13.5 EER 14.0 IEER	

For SI: 1 British thermal unit per hour = 0.2931 W.

a. Chapter 6 contains a complete specification of the referenced test procedure, including the reference year version of the test procedure.

b. Single-phase, air-cooled air conditioners less than 65,000 Btu/h are regulated by NAECA. SEER values are those set by NAECA.

c. Minimum efficiency as of January 1, 2015.

TABLE C403.2.3(2)
MINIMUM EFFICIENCY REQUIREMENTS:
ELECTRICALLY OPERATED UNITARY AND APPLIED HEAT PUMPS

EQUIPMENT TYPE	SIZE CATEGORY	HEATING SECTION TYPE	SUBCATEGORY OR RATING CONDITION	MINIMUM EFFICIENCY Before 1/1/2016	MINIMUM EFFICIENCY As of 1/1/2016	TEST PROCEDURE[a]
Air cooled (cooling mode)	< 65,000 Btu/h[b]	All	Split System	13.0 SEER[c]	14.0 SEER[c]	AHRI 210/240
			Single Package	13.0 SEER[c]	14.0 SEER[c]	
Through-the-wall, air cooled	≤ 30,000 Btu/h[b]	All	Split System	12.0 SEER	12.0 SEER	
			Single Package	12.0 SEER	12.0 SEER	
Single-duct high-velocity air cooled	< 65,000 Btu/h[b]	All	Split System	11.0 SEER	11.0 SEER	
Air cooled (cooling mode)	≥ 65,000 Btu/h and < 135,000 Btu/h	Electric Resistance (or None)	Split System and Single Package	11.0 EER 11.2 IEER	11.0 EER 12.0 IEER	AHRI 340/360
		All other	Split System and Single Package	10.8 EER 11.0 IEER	10.8 EER 11.8 IEER	
	≥ 135,000 Btu/h and < 240,000 Btu/h	Electric Resistance (or None)	Split System and Single Package	10.6 EER 10.7 IEER	10.6 EER 11.6 IEER	
		All other	Split System and Single Package	10.4 EER 10.5 IEER	10.4 EER 11.4 IEER	
	≥ 240,000 Btu/h	Electric Resistance (or None)	Split System and Single Package	9.5 EER 9.6 IEER	9.5 EER 10.6 IEER	
		All other	Split System and Single Package	9.3 EER 9.4 IEER	9.3 EER 9.4 IEER	
Water to Air: Water Loop (cooling mode)	< 17,000 Btu/h	All	86°F entering water	12.2 EER	12.2 EER	ISO 13256-1
	≥ 17,000 Btu/h and < 65,000 Btu/h	All	86°F entering water	13.0 EER	13.0 EER	
	≥ 65,000 Btu/h and < 135,000 Btu/h	All	86°F entering water	13.0 EER	13.0 EER	
Water to Air: Ground Water (cooling mode)	< 135,000 Btu/h	All	59°F entering water	18.0 EER	18.0 EER	ISO 13256-1
Brine to Air: Ground Loop (cooling mode)	< 135,000 Btu/h	All	77°F entering water	14.1 EER	14.1 EER	ISO 13256-1
Water to Water: Water Loop (cooling mode)	< 135,000 Btu/h	All	86°F entering water	10.6 EER	10.6 EER	ISO 13256-2
Water to Water: Ground Water (cooling mode)	< 135,000 Btu/h	All	59°F entering water	16.3 EER	16.3 EER	
Brine to Water: Ground Loop (cooling mode)	< 135,000 Btu/h	All	77°F entering fluid	12.1 EER	12.1 EER	

(continued)

TABLE C403.2.3(2)—continued
MINIMUM EFFICIENCY REQUIREMENTS:
ELECTRICALLY OPERATED UNITARY AND APPLIED HEAT PUMPS

EQUIPMENT TYPE	SIZE CATEGORY	HEATING SECTION TYPE	SUBCATEGORY OR RATING CONDITION	MINIMUM EFFICIENCY Before 1/1/2016	MINIMUM EFFICIENCY As of 1/1/2016	TEST PROCEDURE[a]
Air cooled (heating mode)	< 65,000 Btu/h[b]	—	Split System	7.7 HSPF[c]	8.2 HSPF[c]	AHRI 210/240
		—	Single Package	7.7 HSPF[c]	8.0 HSPF[c]	
Through-the-wall, (air cooled, heating mode)	≤ 30,000 Btu/h[b] (cooling capacity)	—	Split System	7.4 HSPF	7.4 HSPF	
		—	Single Package	7.4 HSPF	7.4 HSPF	
Small-duct high velocity (air cooled, heating mode)	< 65,000 Btu/h[b]	—	Split System	6.8 HSPF	6.8 HSPF	
Air cooled (heating mode)	≥ 65,000 Btu/h and < 135,000 Btu/h (cooling capacity)	—	47°F db/43°F wb outdoor air	3.3 COP	3.3 COP	AHRI 340/360
		—	17°F db/15°F wb outdoor air	2.25 COP	2.25 COP	
	≥ 135,000 Btu/h (cooling capacity)	—	47°F db/43°F wb outdoor air	3.2 COP	3.2 COP	
		—	17°F db/15°F wb outdoor air	2.05 COP	2.05 COP	
Water to Air: Water Loop (heating mode)	< 135,000 Btu/h (cooling capacity)	—	68°F entering water	4.3 COP	4.3 COP	ISO 13256-1
Water to Air: Ground Water (heating mode)	< 135,000 Btu/h (cooling capacity)	—	50°F entering water	3.7 COP	3.7 COP	
Brine to Air: Ground Loop (heating mode)	< 135,000 Btu/h (cooling capacity)	—	32°F entering fluid	3.2 COP	3.2 COP	
Water to Water: Water Loop (heating mode)	< 135,000 Btu/h (cooling capacity)	—	68°F entering water	3.7 COP	3.7 COP	ISO 13256-2
Water to Water: Ground Water (heating mode)	< 135,000 Btu/h (cooling capacity)	—	50°F entering water	3.1 COP	3.1 COP	
Brine to Water: Ground Loop (heating mode)	< 135,000 Btu/h (cooling capacity)	—	32°F entering fluid	2.5 COP	2.5 COP	

For SI: 1 British thermal unit per hour = 0.2931 W, °C = [(°F) - 32]/1.8.

a. Chapter 6 contains a complete specification of the referenced test procedure, including the reference year version of the test procedure.
b. Single-phase, air-cooled air conditioners less than 65,000 Btu/h are regulated by NAECA. SEER values are those set by NAECA.
c. Minimum efficiency as of January 1, 2015.

TABLE C403.2.3(3)
MINIMUM EFFICIENCY REQUIREMENTS:
ELECTRICALLY OPERATED PACKAGED TERMINAL AIR CONDITIONERS,
PACKAGED TERMINAL HEAT PUMPS, SINGLE-PACKAGE VERTICAL AIR CONDITIONERS,
SINGLE VERTICAL HEAT PUMPS, ROOM AIR CONDITIONERS AND ROOM AIR-CONDITIONER HEAT PUMPS

EQUIPMENT TYPE	SIZE CATEGORY (INPUT)	SUBCATEGORY OR RATING CONDITION	MINIMUM EFFICIENCY	TEST PROCEDURE[a]
PTAC (cooling mode) new construction	All Capacities	95°F db outdoor air	14.0 − (0.300 × Cap/1000) EER[c]	AHRI 310/380
PTAC (cooling mode) replacements[b]	All Capacities	95°F db outdoor air	10.9 − (0.213 × Cap/1000) EER	
PTHP (cooling mode) new construction	All Capacities	95°F db outdoor air	14.0 − (0.300 × Cap/1000) EER	
PTHP (cooling mode) replacements[b]	All Capacities	95°F db outdoor air	10.8 − (0.213 × Cap/1000) EER	
PTHP (heating mode) new construction	All Capacities	—	3.2 − (0.026 × Cap/1000) COP	
PTHP (heating mode) replacements[b]	All Capacities	—	2.9 − (0.026 × Cap/1000) COP	
SPVAC (cooling mode)	< 65,000 Btu/h	95°F db/ 75°F wb outdoor air	9.0 EER	AHRI 390
	≥ 65,000 Btu/h and < 135,000 Btu/h	95°F db/ 75°F wb outdoor air	8.9 EER	
	≥ 135,000 Btu/h and < 240,000 Btu/h	95°F db/ 75°F wb outdoor air	8.6 EER	
SPVHP (cooling mode)	< 65,000 Btu/h	95°F db/ 75°F wb outdoor air	9.0 EER	
	≥ 65,000 Btu/h and < 135,000 Btu/h	95°F db/ 75°F wb outdoor air	8.9 EER	
	≥ 135,000 Btu/h and < 240,000 Btu/h	95°F db/ 75°F wb outdoor air	8.6 EER	
SPVHP (heating mode)	< 65,000 Btu/h	47°F db/ 43°F wb outdoor air	3.0 COP	AHRI 390
	≥ 65,000 Btu/h and < 135,000 Btu/h	47°F db/ 43°F wb outdoor air	3.0 COP	
	≥ 135,000 Btu/h and < 240,000 Btu/h	47°F db/ 75°F wb outdoor air	2.9 COP	
Room air conditioners, with louvered sides	< 6,000 Btu/h	—	9.7 SEER	ANSI/ AHAM RAC-1
	≥ 6,000 Btu/h and < 8,000 Btu/h	—	9.7 EER	
	≥ 8,000 Btu/h and < 14,000 Btu/h	—	9.8 EER	
	≥ 14,000 Btu/h and < 20,000 Btu/h	—	9.7 SEER	
	≥ 20,000 Btu/h	—	8.5 EER	
Room air conditioners, without louvered sides	< 8,000 Btu/h	—	9.0 EER	
	≥ 8,000 Btu/h and < 20,000 Btu/h	—	8.5 EER	
	≥ 20,000 Btu/h	—	8.5 EER	
Room air-conditioner heat pumps with louvered sides	< 20,000 Btu/h	—	9.0 EER	
	≥ 20,000 Btu/h	—	8.5 EER	
Room air-conditioner heat pumps without louvered sides	< 14,000 Btu/h	—	8.5 EER	
	≥ 14,000 Btu/h	—	8.0 EER	

(continued)

TABLE C403.2.3(3)—continued
MINIMUM EFFICIENCY REQUIREMENTS:
ELECTRICALLY OPERATED PACKAGED TERMINAL AIR CONDITIONERS,
PACKAGED TERMINAL HEAT PUMPS, SINGLE-PACKAGE VERTICAL AIR CONDITIONERS,
SINGLE VERTICAL HEAT PUMPS, ROOM AIR CONDITIONERS AND ROOM AIR-CONDITIONER HEAT PUMPS

EQUIPMENT TYPE	SIZE CATEGORY (INPUT)	SUBCATEGORY OR RATING CONDITION	MINIMUM EFFICIENCY	TEST PROCEDURE[a]
Room air conditioner casement only	All capacities	—	8.7 EER	ANSI/ AHAM RAC-1
Room air conditioner casement-slider	All capacities	—	9.5 EER	

For SI: 1 British thermal unit per hour = 0.2931 W, °C = [(°F) - 32]/1.8, wb = wet bulb, db = dry bulb.

"Cap" = The rated cooling capacity of the project in Btu/h. Where the unit's capacity is less than 7000 Btu/h, use 7000 Btu/h in the calculation. Where the unit's capacity is greater than 15,000 Btu/h, use 15,000 Btu/h in the calculations.

a. Chapter 6 contains a complete specification of the referenced test procedure, including the referenced year version of the test procedure.
b. Replacement unit shall be factory labeled as follows: "MANUFACTURED FOR REPLACEMENT APPLICATIONS ONLY: NOT TO BE INSTALLED IN NEW CONSTRUCTION PROJECTS." Replacement efficiencies apply only to units with existing sleeves less than 16 inches (406 mm) in height and less than 42 inches (1067 mm) in width.
c. Before January 1, 2015 the minimum efficiency shall be 13.8 - (0.300 x Cap/1000) EER.

TABLE 403.2.3(4)
WARM-AIR FURNACES AND COMBINATION WARM-AIR FURNACES/AIR-CONDITIONING UNITS,
WARM-AIR DUCT FURNACES AND UNIT HEATERS, MINIMUM EFFICIENCY REQUIREMENTS

EQUIPMENT TYPE	SIZE CATEGORY (INPUT)	SUBCATEGORY OR RATING CONDITION	MINIMUM EFFICIENCY[d, e]	TEST PROCEDURE[a]
Warm-air furnaces, gas fired	< 225,000 Btu/h	—	78% AFUE or 80%E_t^c	DOE 10 CFR Part 430 or ANSI Z21.47
	≥ 225,000 Btu/h	Maximum capacity[c]	80%E_t^f	ANSI Z21.47
Warm-air furnaces, oil fired	< 225,000 Btu/h	—	78% AFUE or 80%E_t^c	DOE 10 CFR Part 430 or UL 727
	≥ 225,000 Btu/h	Maximum capacity[b]	81%E_t^g	UL 727
Warm-air duct furnaces, gas fired	All capacities	Maximum capacity[b]	80%E_c	ANSI Z83.8
Warm-air unit heaters, gas fired	All capacities	Maximum capacity[b]	80%E_c	ANSI Z83.8
Warm-air unit heaters, oil fired	All capacities	Maximum capacity[b]	80%E_c	UL 731

For SI: 1 British thermal unit per hour = 0.2931 W.

a. Chapter 6 contains a complete specification of the referenced test procedure, including the referenced year version of the test procedure.
b. Minimum and maximum ratings as provided for and allowed by the unit's controls.
c. Combination units not covered by the National Appliance Energy Conservation Act of 1987 (NAECA) (3-phase power or cooling capacity greater than or equal to 65,000 Btu/h [19 kW]) shall comply with either rating.
d. E_t = Thermal efficiency. See test procedure for detailed discussion.
e. E_c = Combustion efficiency (100% less flue losses). See test procedure for detailed discussion.
f. E_c = Combustion efficiency. Units shall also include an IID, have jackets not exceeding 0.75 percent of the input rating, and have either power venting or a flue damper. A vent damper is an acceptable alternative to a flue damper for those furnaces where combustion air is drawn from the conditioned space.
g. E_t = Thermal efficiency. Units shall also include an IID, have jacket losses not exceeding 0.75 percent of the input rating, and have either power venting or a flue damper. A vent damper is an acceptable alternative to a flue damper for those furnaces where combustion air is drawn from the conditioned space.

COMMERCIAL ENERGY EFFICIENCY

TABLE C403.2.3(5)
MINIMUM EFFICIENCY REQUIREMENTS: GAS- AND OIL-FIRED BOILERS

EQUIPMENT TYPE[a]	SUBCATEGORY OR RATING CONDITION	SIZE CATEGORY (INPUT)	MINIMUM EFFICIENCY[d, e]	TEST PROCEDURE
Boilers, hot water	Gas-fired	< 300,000 Btu/h	80% AFUE	10 CFR Part 430
		≥ 300,000 Btu/h and ≤ 2,500,000 Btu/h[b]	80% E_t	10 CFR Part 431
		> 2,500,000 Btu/h[a]	82% E_c	
	Oil-fired[c]	< 300,000 Btu/h	80% AFUE	10 CFR Part 430
		≥ 300,000 Btu/h and ≤ 2,500,000 Btu/h[b]	82% E_t	10 CFR Part 431
		> 2,500,000 Btu/h[a]	84% E_c	
Boilers, steam	Gas-fired	< 300,000 Btu/h	75% AFUE	10 CFR Part 430
	Gas-fired- all, except natural draft	≥ 300,000 Btu/h and ≤ 2,500,000 Btu/h[b]	79% E_t	10 CFR Part 431
		> 2,500,000 Btu/h[a]	79% E_t	
	Gas-fired-natural draft	≥ 300,000 Btu/h and ≤ 2,500,000 Btu/h[b]	77% E_t	
		> 2,500,000 Btu/h[a]	77% E_t	
	Oil-fired[c]	< 300,000 Btu/h	80% AFUE	10 CFR Part 430
		≥ 300,000 Btu/h and ≤ 2,500,000 Btu/h[b]	81% E_t	10 CFR Part 431
		> 2,500,000 Btu/h[a]	81% E_t	

For SI: 1 British thermal unit per hour = 0.2931 W.

a. These requirements apply to boilers with rated input of 8,000,000 Btu/h or less that are not packaged boilers and to all packaged boilers. Minimum efficiency requirements for boilers cover all capacities of packaged boilers.
b. Maximum capacity – minimum and maximum ratings as provided for and allowed by the unit's controls.
c. Includes oil-fired (residual).
d. E_c = Combustion efficiency (100 percent less flue losses).
e. E_t = Thermal efficiency. See referenced standard for detailed information.

TABLE C403.2.3(6)
MINIMUM EFFICIENCY REQUIREMENTS:
CONDENSING UNITS, ELECTRICALLY OPERATED

EQUIPMENT TYPE	SIZE CATEGORY	MINIMUM EFFICIENCY[b]	TEST PROCEDURE[a]
Condensing units, air cooled	≥ 135,000 Btu/h	10.1 EER 11.2 IPLV	AHRI 365
Condensing units, water or evaporatively cooled	≥ 135,000 Btu/h	13.1 EER 13.1 IPLV	

For SI: 1 British thermal unit per hour = 0.2931 W.

a. Chapter 6 contains a complete specification of the referenced test procedure, including the referenced year version of the test procedure.
b. IPLVs are only applicable to equipment with capacity modulation.

TABLE C403.2.3(7)
WATER CHILLING PACKAGES – EFFICIENCY REQUIREMENTS[a, b, d]

EQUIPMENT TYPE	SIZE CATEGORY	UNITS	BEFORE 1/1/2015 Path A	BEFORE 1/1/2015 Path B	AS OF 1/1/2015 Path A	AS OF 1/1/2015 Path B	TEST PROCEDURE[c]
Air-cooled chillers	< 150 Tons	EER (Btu/W)	≥ 9.562 FL	NA[c]	≥ 10.100 FL	≥ 9.700 FL	
			≥ 12.500 IPLV		≥ 13.700 IPLV	≥ 15.800 IPLV	
	≥ 150 Tons		≥ 9.562 FL	NA[c]	≥ 10.100 FL	≥ 9.700 FL	
			≥ 12.500 IPLV		≥ 14.000 IPLV	≥ 16.100 IPLV	
Air cooled without condenser, electrically operated	All capacities	EER (Btu/W)	Air-cooled chillers without condenser shall be rated with matching condensers and complying with air-cooled chiller efficiency requirements.				
Water cooled, electrically operated positive displacement	< 75 Tons	kW/ton	≤ 0.780 FL	≤ 0.800 FL	≤ 0.750 FL	≤ 0.780 FL	AHRI 550/ 590
			≤ 0.630 IPLV	≤ 0.600 IPLV	≤ 0.600 IPLV	≤ 0.500 IPLV	
	≥ 75 tons and < 150 tons		≤ 0.775 FL	≤ 0.790 FL	≤ 0.720 FL	≤ 0.750 FL	
			≤ 0.615 IPLV	≤ 0.586 IPLV	≤ 0.560 IPLV	≤ 0.490 IPLV	
	≥ 150 tons and < 300 tons		≤ 0.680 FL	≤ 0.718 FL	≤ 0.660 FL	≤ 0.680 FL	
			≤ 0.580 IPLV	≤ 0.540 IPLV	≤ 0.540 IPLV	≤ 0.440 IPLV	
	≥ 300 tons and < 600 tons		≤ 0.620 FL	≤ 0.639 FL	≤ 0.610 FL	≤ 0.625 FL	
			≤ 0.540 IPLV	≤ 0.490 IPLV	≤ 0.520 IPLV	≤ 0.410 IPLV	
	≥ 600 tons		≤ 0.620 FL	≤ 0.639 FL	≤ 0.560 FL	≤ 0.585 FL	
			≤ 0.540 IPLV	≤ 0.490 IPLV	≤ 0.500 IPLV	≤ 0.380 IPLV	
Water cooled, electrically operated centrifugal	< 150 Tons	kW/ton	≤ 0.634 FL	≤ 0.639 FL	≤ 0.610 FL	≤ 0.695 FL	
			≤ 0.596 IPLV	≤ 0.450 IPLV	≤ 0.550 IPLV	≤ 0.440 IPLV	
	≥ 150 tons and < 300 tons		≤ 0.634 FL	≤ 0.639 FL	≤ 0.610 FL	≤ 0.635 FL	
			≤ 0.596 IPLV	≤ 0.450 IPLV	≤ 0.550 IPLV	≤ 0.400 IPLV	
	≥ 300 tons and < 400 tons		≤ 0.576 FL	≤ 0.600 FL	≤ 0.560 FL	≤ 0.595 FL	
			≤ 0.549 IPLV	≤ 0.400 IPLV	≤ 0.520 IPLV	≤ 0.390 IPLV	
	≥ 400 tons and < 600 tons		≤ 0.576 FL	≤ 0.600 FL	≤ 0.560 FL	≤ 0.585 FL	
			≤ 0.549 IPLV	≤ 0.400 IPLV	≤ 0.500 IPLV	≤ 0.380 IPLV	
	≥ 600 Tons		≤ 0.570 FL	≤ 0.590 FL	≤ 0.560 FL	≤ 0.585 FL	
			≤ 0.539 IPLV	≤ 0.400 IPLV	≤ 0.500 IPLV	≤ 0.380 IPLV	
Air cooled, absorption, single effect	All capacities	COP	≥ 0.600 FL	NA[c]	≥ 0.600 FL	NA[c]	AHRI 560
Water cooled absorption, single effect	All capacities	COP	≥ 0.700 FL	NA[c]	≥ 0.700 FL	NA[c]	
Absorption, double effect, indirect fired	All capacities	COP	≥ 1.000 FL	NA[c]	≥ 1.000 FL	NA[c]	
			≥ 1.050 IPLV		≥ 1.050 IPLV		
Absorption double effect direct fired	All capacities	COP	≥ 1.000 FL	NA[c]	≥ 1.000 FL	NA[c]	
			≥ 1.000 IPLV		≥ 1.050 IPLV		

a. The requirements for centrifugal chiller shall be adjusted for nonstandard rating conditions in accordance with Section C403.2.3.1 and are only applicable for the range of conditions listed in Section C403.2.3.1. The requirements for air-cooled, water-cooled positive displacement and absorption chillers are at standard rating conditions defined in the reference test procedure.
b. Both the full-load and IPLV requirements shall be met or exceeded to comply with this standard. Where there is a Path B, compliance can be with either Path A or Path B for any application.
c. NA means the requirements are not applicable for Path B and only Path A can be used for compliance.
d. FL represents the full-load performance requirements and IPLV the part-load performance requirements.

TABLE C403.2.3(8)
MINIMUM EFFICIENCY REQUIREMENTS:
HEAT REJECTION EQUIPMENT

EQUIPMENT TYPE[a]	TOTAL SYSTEM HEAT REJECTION CAPACITY AT RATED CONDITIONS	SUBCATEGORY OR RATING CONDITION[i]	PERFORMANCE REQUIRED[b, c, d, g, h]	TEST PROCEDURE[e, f]
Propeller or axial fan open-circuit cooling towers	All	95°F entering water 85°F leaving water 75°F entering wb	≥ 40.2 gpm/hp	CTI ATC-105 and CTI STD-201
Centrifugal fan open-circuit cooling towers	All	95°F entering water 85°F leaving water 75°F entering wb	≥ 20.0 gpm/hp	CTI ATC-105 and CTI STD-201
Propeller or axial fan closed-circuit cooling towers	All	102°F entering water 90°F leaving water 75°F entering wb	≥ 14.0 gpm/hp	CTI ATC-105S and CTI STD-201
Centrifugal fan closed-circuit cooling towers	All	102°F entering water 90°F leaving water 75°F entering wb	≥ 7.0 gpm/hp	CTI ATC-105S and CTI STD-201
Propeller or axial fan evaporative condensers	All	Ammonia Test Fluid 140°F entering gas temperature 96.3°F condensing temperature 75°F entering wb	≥ 134,000 Btu/h·hp	CTI ATC-106
Centrifugal fan evaporative condensers	All	Ammonia Test Fluid 140°F entering gas temperature 96.3°F condensing temperature 75°F entering wb	≥ 110,000 Btu/h·hp	CTI ATC-106
Propeller or axial fan evaporative condensers	All	R-507A Test Fluid 165°F entering gas temperature 105°F condensing temperature 75°F entering wb	≥ 157,000 Btu/h·hp	CTI ATC-106
Centrifugal fan evaporative condensers	All	R-507A Test Fluid 165°F entering gas temperature 105°F condensing temperature 75°F entering wb	≥ 135,000 Btu/h·hp	CTI ATC-106
Air-cooled condensers	All	125°F Condensing Temperature 190°F Entering Gas Temperature 15°F subcooling 95°F entering db	≥ 176,000 Btu/h·hp	AHRI 460

For SI: °C = [(°F)-32]/1.8, L/s · kW = (gpm/hp)/(11.83), COP = (Btu/h · hp)/(2550.7),
db = dry bulb temperature, °F, wb = wet bulb temperature, °F.

a. The efficiencies and test procedures for both open- and closed-circuit cooling towers are not applicable to hybrid cooling towers that contain a combination of wet and dry heat exchange sections.
b. For purposes of this table, open circuit cooling tower performance is defined as the water flow rating of the tower at the thermal rating condition listed in Table 403.2.3(8) divided by the fan nameplate-rated motor power.
c. For purposes of this table, closed-circuit cooling tower performance is defined as the water flow rating of the tower at the thermal rating condition listed in Table 403.2.3(8) divided by the sum of the fan nameplate-rated motor power and the spray pump nameplate-rated motor power.
d. For purposes of this table, air-cooled condenser performance is defined as the heat rejected from the refrigerant divided by the fan nameplate-rated motor power.
e. Chapter 6 contains a complete specification of the referenced test procedure, including the referenced year version of the test procedure. The certification requirements do not apply to field-erected cooling towers.
f. Where a certification program exists for a covered product and it includes provisions for verification and challenge of equipment efficiency ratings, then the product shall be listed in the certification program; or, where a certification program exists for a covered product, and it includes provisions for verification and challenge of equipment efficiency ratings, but the product is not listed in the existing certification program, the ratings shall be verified by an independent laboratory test report.
g. Cooling towers shall comply with the minimum efficiency listed in the table for that specific type of tower with the capacity effect of any project-specific accessories and/or options included in the capacity of the cooling tower
h. For purposes of this table, evaporative condenser performance is defined as the heat rejected at the specified rating condition in the table divided by the sum of the fan motor nameplate power and the integral spray pump nameplate power
i. Requirements for evaporative condensers are listed with ammonia (R-717) and R-507A as test fluids in the table. Evaporative condensers intended for use with halocarbon refrigerants other than R-507A shall meet the minimum efficiency requirements listed in this table with R-507A as the test fluid.

TABLE C403.2.3(9)
MINIMUM EFFICIENCY AIR CONDITIONERS AND CONDENSING UNITS SERVING COMPUTER ROOMS

EQUIPMENT TYPE	NET SENSIBLE COOLING CAPACITY[a]	MINIMUM SCOP-127[b] EFFICIENCY DOWNFLOW UNITS / UPFLOW UNITS	TEST PROCEDURE
Air conditioners, air cooled	< 65,000 Btu/h	2.20 / 2.09	ANSI/ASHRAE 127
	≥ 65,000 Btu/h and < 240,000 Btu/h	2.10 / 1.99	
	≥ 240,000 Btu/h	1.90 / 1.79	
Air conditioners, water cooled	< 65,000 Btu/h	2.60 / 2.49	
	≥ 65,000 Btu/h and < 240,000 Btu/h	2.50 / 2.39	
	≥ 240,000 Btu/h	2.40 / 2.29	
Air conditioners, water cooled with fluid economizer	< 65,000 Btu/h	2.55 / 2.44	
	≥ 65,000 Btu/h and < 240,000 Btu/h	2.45 / 2.34	
	≥ 240,000 Btu/h	2.35 / 2.24	
Air conditioners, glycol cooled (rated at 40% propylene glycol)	< 65,000 Btu/h	2.50 / 2.39	
	≥ 65,000 Btu/h and < 240,000 Btu/h	2.15 / 2.04	
	≥ 240,000 Btu/h	2.10 / 1.99	
Air conditioners, glycol cooled (rated at 40% propylene glycol) with fluid economizer	< 65,000 Btu/h	2.45 / 2.34	
	≥ 65,000 Btu/h and < 240,000 Btu/h	2.10 / 1.99	
	≥ 240,000 Btu/h	2.05 / 1.94	

For SI: 1 British thermal unit per hour = 0.2931 W.

a. Net sensible cooling capacity: the total gross cooling capacity less the latent cooling less the energy to the air movement system. (Total Gross – latent – Fan Power).

b. Sensible coefficient of performance (SCOP-127): a ratio calculated by dividing the net sensible cooling capacity in watts by the total power input in watts (excluding reheaters and humidifiers) at conditions defined in ASHRAE Standard 127. The net sensible cooling capacity is the gross sensible capacity minus the energy dissipated into the cooled space by the fan system.

TABLE C403.2.3(10)
HEAT TRANSFER EQUIPMENT

EQUIPMENT TYPE	SUBCATEGORY	MINIMUM EFFICIENCY	TEST PROCEDURE[a]
Liquid-to-liquid heat exchangers	Plate type	NR	AHRI 400

NR = No Requirement.

a. Chapter 6 contains a complete specification of the referenced test procedure, including the referenced year version of the test procedure.

C403.2.3.1 Water-cooled centrifugal chilling packages. Equipment not designed for operation at AHRI Standard 550/590 test conditions of 44°F (7°C) leaving chilled-water temperature and 2.4 gpm/ton evaporator fluid flow and 85°F (29°C) entering condenser water temperature with 3 gpm/ton (0.054 l/s · kW) condenser water flow shall have maximum full-load kW/ton (FL) and part-load ratings requirements adjusted using Equations 4-6 and 4-7.

$FL_{adj} = FL/K_{adj}$ **(Equation 4-6)**

$PLV_{adj} = IPLV/K_{adj}$ **(Equation 4-7)**

where:

K_{adj} = $A \times B$

FL = Full-load kW/ton value as specified in Table C403.2.3(7).

FL_{adj} = Maximum full-load kW/ton rating, adjusted for nonstandard conditions.

$IPLV$ = Value as specified in Table C403.2.3(7).

PLV_{adj} = Maximum NPLV rating, adjusted for nonstandard conditions.

A = $0.00000014592 \cdot (LIFT)^4 - 0.0000346496 \cdot (LIFT)^3 + 0.00314196 \cdot (LIFT)^2 - 0.147199 \cdot (LIFT) + 3.9302$

B = $0.0015 \cdot L_{vg}E_{vap} + 0.934$

$LIFT$ = $L_{vg}Cond - L_{vg}E_{vap}$

$L_{vg}Cond$ = Full-load condenser leaving fluid temperature (°F).

$L_{vg}E_{vap}$ = Full-load evaporator leaving temperature (°F).

The FL_{adj} and PLV_{adj} values are only applicable for centrifugal chillers meeting all of the following full-load design ranges:

1. Minimum evaporator leaving temperature: 36°F.

2. Maximum condenser leaving temperature: 115°F.

3. $20°F \leq LIFT \leq 80°F$.

C403.2.3.2 Positive displacement (air- and water-cooled) chilling packages. Equipment with a leaving fluid temperature higher than 32°F (0°C) and water-cooled positive displacement chilling packages with a condenser leaving fluid temperature below 115°F (46°C) shall meet the requirements of Table C403.2.3(7) when tested or certified with water at standard rating conditions, in accordance with the referenced test procedure.

C403.2.4 HVAC system controls. Each heating and cooling system shall be provided with thermostatic controls as specified in Section C403.2.4.1, C403.2.4.1.3, C403.2.4.2, C403.2.4.3, C403.3.1, C403.4, C403.4.1 or C403.4.4.

C403.2.4.1 Thermostatic controls. The supply of heating and cooling energy to each *zone* shall be controlled by individual thermostatic controls capable of responding to temperature within the *zone*. Where humidification or dehumidification or both is provided, at least one humidity control device shall be provided for each humidity control system.

> **Exception:** Independent perimeter systems that are designed to offset only building envelope heat losses, gains or both serving one or more perimeter *zone*s also served by an interior system provided:
>
> 1. The perimeter system includes at least one thermostatic control *zone* for each building exposure having exterior walls facing only one orientation (within +/-45 degrees) (0.8 rad) for more than 50 contiguous feet (15 240 mm); and
> 2. The perimeter system heating and cooling supply is controlled by thermostats located within the *zones* served by the system.

C403.2.4.1.1 Heat pump supplementary heat. Heat pumps having supplementary electric resistance heat shall have controls that, except during defrost, prevent supplementary heat operation where the heat pump can provide the heating load.

C403.2.4.1.2 Deadband. Where used to control both heating and cooling, *zone* thermostatic controls shall be capable of providing a temperature range or deadband of at least 5°F (2.8°C) within which the supply of heating and cooling energy to the *zone* is capable of being shut off or reduced to a minimum.

> **Exceptions:**
>
> 1. Thermostats requiring manual changeover between heating and cooling modes.
> 2. Occupancies or applications requiring precision in indoor temperature control as *approved* by the *code official*.

C403.2.4.1.3 Set point overlap restriction. Where a *zone* has a separate heating and a separate cooling thermostatic control located within the *zone*, a limit switch, mechanical stop or direct digital control system with software programming shall be provided with the capability to prevent the heating set point from exceeding the cooling set point and to maintain a deadband in accordance with Section C403.2.4.1.2.

C403.2.4.2 Off-hour controls. Each *zone* shall be provided with thermostatic setback controls that are controlled by either an automatic time clock or programmable control system.

> **Exceptions:**
>
> 1. *Zones* that will be operated continuously.
> 2. *Zones* with a full HVAC load demand not exceeding 6,800 Btu/h (2 kW) and having a readily accessible manual shutoff switch.

C403.2.4.2.1 Thermostatic setback capabilities. Thermostatic setback controls shall have the capability to set back or temporarily operate the system to maintain *zone* temperatures down to 55°F (13°C) or up to 85°F (29°C).

C403.2.4.2.2 Automatic setback and shutdown capabilities. Automatic time clock or programmable controls shall be capable of starting and stopping the system for seven different daily schedules per week and retaining their programming and time setting during a loss of power for at least 10 hours. Additionally, the controls shall have a manual override that allows temporary operation of the system for up to 2 hours; a manually operated timer capable of being adjusted to operate the system for up to 2 hours; or an occupancy sensor.

C403.2.4.2.3 Automatic start capabilities. Automatic start controls shall be provided for each HVAC system. The controls shall be capable of automatically adjusting the daily start time of the HVAC system in order to bring each space to the desired occupied temperature immediately prior to scheduled occupancy.

C403.2.4.3 Shutoff dampers. Outdoor air intake and exhaust openings and stairway and shaft vents shall be provided with Class I motorized dampers. The dampers shall have an air leakage rate not greater than 4 cfm/ft^2 (20.3 L/s · m^2) of damper surface area at 1.0 inch water gauge (249 Pa) and shall be labeled by an approved agency when tested in accordance with AMCA 500D for such purpose.

Outdoor air intake and exhaust dampers shall be installed with automatic controls configured to close when the systems or spaces served are not in use or during unoccupied period warm-up and setback operation, unless the systems served require outdoor or exhaust air in accordance with the *International Mechanical Code* or the dampers are opened to provide intentional economizer cooling.

Stairway and shaft vent dampers shall be installed with automatic controls configured to open upon the activation of any fire alarm initiating device of the building's fire alarm system or the interruption of power to the damper.

> **Exception:** Gravity (nonmotorized) dampers shall be permitted to be used as follows:
>
> 1. In buildings less than three stories in height above grade plane.
> 2. In buildings of any height located in *Climate Zones* 1, 2 or 3.
> 3. Where the design exhaust capacity is not greater than 300 cfm (142 L/s).

Gravity (nonmotorized) dampers shall have an air leakage rate not greater than 20 cfm/ft^2 (101.6 L/s · m^2) where not less than 24 inches (610 mm) in either dimension and 40 cfm/ft^2 (203.2 L/s · m^2) where less than 24 inches (610 mm) in either dimension. The rate of air leakage shall be determined at 1.0 inch water gauge (249 Pa) when tested in accordance with AMCA 500D for such purpose. The dampers shall be labeled by an approved agency.

C403.2.4.4 Zone isolation. HVAC systems serving *zones* that are over 25,000 square feet (2323 m^2) in floor area or that span more than one floor and are designed to operate or be occupied nonsimultaneously shall be divided into isolation areas. Each isolation area shall be equipped with isolation devices and controls configured to automatically shut off the supply of conditioned air and outdoor air to and exhaust air from the isolation area. Each isolation area shall be controlled independently by a device meeting the requirements of Section C403.2.4.2.2. Central systems and plants shall be provided with controls and devices that will allow system and equipment operation for any length of time while serving only the smallest isolation area served by the system or plant.

> **Exceptions:**
>
> 1. Exhaust air and outdoor air connections to isolation areas where the fan system to which they connect is not greater than 5,000 cfm (2360 L/s).
> 2. Exhaust airflow from a single isolation area of less than 10 percent of the design airflow of the exhaust system to which it connects.
> 3. Isolation areas intended to operate continuously or intended to be inoperative only when all other isolation areas in a *zone* are inoperative.

C403.2.4.5 Snow- and ice-melt system controls. Snow- and ice-melting systems shall include automatic controls capable of shutting off the system when the pavement temperature is above 50°F (10°C) and no precipitation is falling and an automatic or manual control that will allow shutoff when the outdoor temperature is above 40°F (4°C).

C403.2.4.6 Freeze protection system controls. Freeze protection systems, such as heat tracing of outdoor piping and heat exchangers, including self-regulating heat tracing, shall include automatic controls configured to shut off the systems when outdoor air temperatures are above 40°F (4°C) or when the conditions of the protected fluid will prevent freezing.

C403.2.4.7 Economizer fault detection and diagnostics (FDD). Air-cooled unitary direct-expansion units listed in Tables C403.2.3(1) through C403.2.3(3) and variable refrigerant flow (VRF) units that are equipped with an economizer in accordance with Section C403.3 shall include a fault detection and diagnostics (FDD) system complying with the following:

1. The following temperature sensors shall be permanently installed to monitor system operation:
    1.1. Outside air.
    1.2. Supply air.
    1.3. Return air.
2. Temperature sensors shall have an accuracy of ±2°F (1.1°C) over the range of 40°F to 80°F (4°C to 26.7°C).
3. Refrigerant pressure sensors, where used, shall have an accuracy of ±3 percent of full scale.
4. The unit controller shall be capable of providing system status by indicating the following:
    4.1. Free cooling available.
    4.2. Economizer enabled.
    4.3. Compressor enabled.
    4.4. Heating enabled.
    4.5. Mixed air low limit cycle active.
    4.6. The current value of each sensor.
5. The unit controller shall be capable of manually initiating each operating mode so that the operation of compressors, economizers, fans and the heating system can be independently tested and verified.
6. The unit shall be capable of reporting faults to a fault management application accessible by day-to-day operating or service personnel, or annunciated locally on zone thermostats.
7. The FDD system shall be capable of detecting the following faults:
    7.1. Air temperature sensor failure/fault.
    7.2. Not economizing when the unit should be economizing.
    7.3. Economizing when the unit should not be economizing.
    7.4. Damper not modulating.
    7.5. Excess outdoor air.

C403.2.5 Hot water boiler outdoor temperature setback control. Hot water boilers that supply heat to the building through one- or two-pipe heating systems shall have an outdoor setback control that lowers the boiler water temperature based on the outdoor temperature.

C403.2.6 Ventilation. Ventilation, either natural or mechanical, shall be provided in accordance with Chapter 4 of the

International Mechanical Code. Where mechanical ventilation is provided, the system shall provide the capability to reduce the outdoor air supply to the minimum required by Chapter 4 of the *International Mechanical Code.*

C403.2.6.1 Demand controlled ventilation. Demand control ventilation (DCV) shall be provided for spaces larger than 500 square feet (46.5 m^2) and with an average occupant load of 25 people per 1,000 square feet (93 m^2) of floor area (as established in Table 403.3.1.1 of the *International Mechanical Code*) and served by systems with one or more of the following:

1. An air-side economizer.
2. Automatic modulating control of the outdoor air damper.
3. A design outdoor airflow greater than 3,000 cfm (1416 L/s).

Exception: Demand control ventilation is not required for systems and spaces as follows:

1. Systems with energy recovery complying with Section C403.2.7.
2. Multiple-*zone* systems without direct digital control of individual *zones* communicating with a central control panel.
3. Systems with a design outdoor airflow less than 1,200 cfm (566 L/s).
4. Spaces where the supply airflow rate minus any makeup or outgoing transfer air requirement is less than 1,200 cfm (566 L/s).
5. Ventilation provided for process loads only.

C403.2.6.2 Enclosed parking garage ventilation controls. Enclosed parking garages used for storing or handling automobiles operating under their own power shall employ contamination-sensing devices and automatic controls configured to stage fans or modulate fan average airflow rates to 50 percent or less of design capacity, or intermittently operate fans less than 20 percent of the occupied time or as required to maintain acceptable contaminant levels in accordance with *International Mechanical Code* provisions. Failure of contamination sensing devices shall cause the exhaust fans to operate continuously at design airflow.

Exceptions:

1. Garages with a total exhaust capacity less than 22,500 cfm (10 620 L/s) with ventilation systems that do not utilize heating or mechanical cooling.
2. Garages that have a garage area to ventilation system motor nameplate power ratio that exceeds 1125 cfm/hp (710 L/s/kW) and do not utilize heating or mechanical cooling.

C403.2.7 Energy recovery ventilation systems. Where the supply airflow rate of a fan system exceeds the values specified in Tables C403.2.7(1) and C403.2.7(2), the system shall include an energy recovery system. The energy recovery system shall have the capability to provide a change in the enthalpy of the outdoor air supply of not less than 50 percent of the difference between the outdoor air

TABLE C403.2.7(1)
ENERGY RECOVERY REQUIREMENT
(Ventilation systems operating less than 8,000 hours per year)

CLIMATE ZONE	PERCENT (%) OUTDOOR AIR AT FULL DESIGN AIRFLOW RATE							
	≥ 10% and < 20%	≥ 20% and < 30%	≥ 30% and < 40%	≥ 40% and < 50%	≥ 50% and < 60%	≥ 60% and < 70%	≥ 70% and < 80%	≥ 80%
	DESIGN SUPPLY FAN AIRFLOW RATE (cfm)							
3B, 3C, 4B, 4C, 5B	NR	NR	NR	NR	NR	NR	NR	NR
1B, 2B, 5C	NR	NR	NR	NR	≥ 26,000	≥ 12,000	≥ 5,000	≥ 4,000
6B	≥ 28,000	≥ 26,5000	≥ 11,000	≥ 5,500	≥ 4,500	≥ 3,500	≥ 2,500	≥ 1,500
1A, 2A, 3A, 4A, 5A, 6A	≥ 26,000	≥ 16,000	≥ 5,500	≥ 4,500	≥ 3,500	≥ 2,000	≥ 1,000	> 0
7, 8	≥ 4,500	≥ 4,000	≥ 2,500	≥ 1,000	> 0	> 0	> 0	> 0

For SI: 1 cfm = 0.4719 L/s.
NR = Not Required.

TABLE C403.2.7(2)
ENERGY RECOVERY REQUIREMENT
(Ventilation systems operating not less than 8,000 hours per year)

CLIMATE ZONE	PERCENT (%) OUTDOOR AIR AT FULL DESIGN AIRFLOW RATE							
	≥ 10% and < 20%	≥ 20% and < 30%	≥ 30% and < 40%	≥ 40% and < 50%	≥ 50% and < 60%	≥ 60% and < 70%	≥ 70% and < 80%	≥ 80%
	Design Supply Fan Airflow Rate (cfm)							
3C	NR	NR	NR	NR	NR	NR	NR	NR
1B, 2B, 3B, 4C, 5C	NR	≥ 19,500	≥ 9,000	≥ 5,000	≥ 4,000	≥ 3,000	≥ 1,500	> 0
1A, 2A, 3A, 4B, 5B	≥ 2,500	≥ 2,000	≥ 1,000	≥ 500	> 0	> 0	> 0	> 0
4A, 5A, 6A, 6B, 7, 8	> 0	> 0	> 0	> 0	> 0	> 0	> 0	> 0

For SI: 1 cfm = 0.4719 L/s.
NR = Not required

and return air enthalpies, at design conditions. Where an air economizer is required, the energy recovery system shall include a bypass or controls which permit operation of the economizer as required by Section C403.3.

Exception: An energy recovery ventilation system shall not be required in any of the following conditions:

1. Where energy recovery systems are prohibited by the *International Mechanical Code*.
2. Laboratory fume hood systems that include at least one of the following features:
   2.1. Variable-air-volume hood exhaust and room supply systems capable of reducing exhaust and makeup air volume to 50 percent or less of design values.
   2.2. Direct makeup (auxiliary) air supply equal to at least 75 percent of the exhaust rate, heated not warmer than 2°F (1.1°C) above room setpoint, cooled to not cooler than 3°F (1.7°C) below room setpoint, no humidification added, and no simultaneous heating and cooling used for dehumidification control.
3. Systems serving spaces that are heated to less than 60°F (15.5°C) and are not cooled.
4. Where more than 60 percent of the outdoor heating energy is provided from site-recovered or site solar energy.
5. Heating energy recovery in *Climate Zones* 1 and 2.
6. Cooling energy recovery in *Climate Zones* 3C, 4C, 5B, 5C, 6B, 7 and 8.
7. Systems requiring dehumidification that employ energy recovery in series with the cooling coil.
8. Where the largest source of air exhausted at a single location at the building exterior is less than 75 percent of the design *outdoor air* flow rate.
9. Systems expected to operate less than 20 hours per week at the *outdoor air* percentage covered by Table C403.2.7(1).
10. Systems exhausting toxic, flammable, paint or corrosive fumes or dust.
11. Commercial kitchen hoods used for collecting and removing grease vapors and smoke.

C403.2.8 Kitchen exhaust systems. Replacement air introduced directly into the exhaust hood cavity shall not be greater than 10 percent of the hood exhaust airflow rate. Conditioned supply air delivered to any space shall not exceed the greater of the following:

1. The ventilation rate required to meet the space heating or cooling load.
2. The hood exhaust flow minus the available transfer air from adjacent space where available transfer air is considered that portion of outdoor ventilation air not required to satisfy other exhaust needs, such as restrooms, and not required to maintain pressurization of adjacent spaces.

Where total kitchen hood exhaust airflow rate is greater than 5,000 cfm (2360 L/s), each hood shall be a factory-built commercial exhaust hood listed by a nationally recognized testing laboratory in compliance with UL 710. Each hood shall have a maximum exhaust rate as specified in Table C403.2.8 and shall comply with one of the following:

1. Not less than 50 percent of all replacement air shall be transfer air that would otherwise be exhausted.
2. Demand ventilation systems on not less than 75 percent of the exhaust air that are capable of not less than a 50-percent reduction in exhaust and replacement air system airflow rates, including controls necessary to modulate airflow in response to appliance operation and to maintain full capture and containment of smoke, effluent and combustion products during cooking and idle.
3. Listed energy recovery devices with a sensible heat recovery effectiveness of not less than 40 percent on not less than 50 percent of the total exhaust airflow.

Where a single hood, or hood section, is installed over appliances with different duty ratings, the maximum allowable flow rate for the hood or hood section shall be based on the requirements for the highest appliance duty rating under the hood or hood section.

Exception: Where not less than 75 percent of all the replacement air is transfer air that would otherwise be exhausted

TABLE C403.2.8
MAXIMUM NET EXHAUST FLOW RATE,
CFM PER LINEAR FOOT OF HOOD LENGTH

TYPE OF HOOD	LIGHT-DUTY EQUIPMENT	MEDIUM-DUTY EQUIPMENT	HEAVY-DUTY EQUIPMENT	EXTRA-HEAVY-DUTY EQUIPMENT
Wall-mounted canopy	140	210	280	385
Single island	280	350	420	490
Double island (per side)	175	210	280	385
Eyebrow	175	175	NA	NA
Backshelf/Pass-over	210	210	280	NA

For SI: 1 cfm = 0.4719 L/s; 1 foot = 305 mm.
NA = Not Allowed.

C403.2.9 Duct and plenum insulation and sealing. Supply and return air ducts and plenums shall be insulated with a minimum of R-6 insulation where located in unconditioned spaces and where located outside the building with a minimum of R-8 insulation in *Climate Zones* 1 through 4 and a minimum of R-12 insulation in *Climate Zones* 5 through 8. Where located within a building envelope assembly, the duct or plenum shall be separated from the building exterior or unconditioned or exempt spaces by a minimum of R-8 insulation in *Climate Zones* 1 through 4 and a minimum of R-12 insulation in *Climate Zones* 5 through 8.

Exceptions:

1. Where located within equipment.
2. Where the design temperature difference between the interior and exterior of the duct or plenum is not greater than 15°F (8°C).

Ducts, air handlers and filter boxes shall be sealed. Joints and seams shall comply with Section 603.9 of the *International Mechanical Code*.

C403.2.9.1 Duct construction. Ductwork shall be constructed and erected in accordance with the *International Mechanical Code*.

C403.2.9.1.1 Low-pressure duct systems. Longitudinal and transverse joints, seams and connections of supply and return ducts operating at a static pressure less than or equal to 2 inches water gauge (w.g.) (498 Pa) shall be securely fastened and sealed with welds, gaskets, mastics (adhesives), mastic-plus-embedded-fabric systems or tapes installed in accordance with the manufacturer's instructions. Pressure classifications specific to the duct system shall be clearly indicated on the construction documents in accordance with the *International Mechanical Code*.

Exception: Locking-type longitudinal joints and seams, other than the snap-lock and button-lock types, need not be sealed as specified in this section.

C403.2.9.1.2 Medium-pressure duct systems. Ducts and plenums designed to operate at a static pressure greater than 2 inches water gauge (w.g.) (498 Pa) but less than 3 inches w.g. (747 Pa) shall be insulated and sealed in accordance with Section C403.2.9. Pressure classifications specific to the duct system shall be clearly indicated on the construction documents in accordance with the *International Mechanical Code*.

C403.2.9.1.3 High-pressure duct systems. Ducts and plenums designed to operate at static pressures greater than 3 inches water gauge (747 Pa) shall be insulated and sealed in accordance with Section C403.2.9. In addition, ducts and plenums shall be leak tested in accordance with the SMACNA *HVAC Air Duct Leakage Test Manual* and shown to have a rate of air leakage (CL) less than or equal to 4.0 as determined in accordance with Equation 4-8.

$$CL = F/P^{0.65} \quad \text{(Equation 4-8)}$$

where:

F = The measured leakage rate in cfm per 100 square feet of duct surface.

P = The static pressure of the test.

Documentation shall be furnished by the designer demonstrating that representative sections totaling at least 25 percent of the duct area have been tested and that all tested sections comply with the requirements of this section.

C403.2.10 Piping insulation. Piping serving as part of a heating or cooling system shall be thermally insulated in accordance with Table C403.2.10.

Exceptions:

1. Factory-installed piping within HVAC equipment tested and rated in accordance with a test procedure referenced by this code.
2. Factory-installed piping within room fan-coils and unit ventilators tested and rated according to AHRI 440 (except that the sampling and variation provisions of Section 6.5 shall not apply) and AHRI 840, respectively.
3. Piping that conveys fluids that have a design operating temperature range between 60°F (15°C) and 105°F (41°C).
4. Piping that conveys fluids that have not been heated or cooled through the use of fossil fuels or electric power.
5. Strainers, control valves, and balancing valves associated with piping 1 inch (25 mm) or less in diameter.
6. Direct buried piping that conveys fluids at or below 60°F (15°C).

C403.2.10.1 Protection of piping insulation. Piping insulation exposed to the weather shall be protected from damage, including that due to sunlight, moisture, equipment maintenance and wind, and shall provide shielding from solar radiation that can cause degradation of the material. Adhesive tape shall not be permitted.

C403.2.11 Mechanical systems commissioning and completion requirements. Mechanical systems shall be commissioned and completed in accordance with Section C408.2.

C403.2.12 Air system design and control. Each HVAC system having a total fan system motor nameplate horsepower (hp) exceeding 5 hp (3.7 kW) shall comply with the provisions of Sections C403.2.12.1 through C403.2.12.3.

C403.2.12.1 Allowable fan motor horsepower. Each HVAC system at fan system design conditions shall not exceed the allowable *fan system motor nameplate hp* (Option 1) or *fan system bhp* (Option 2) as shown in Table C403.2.12.1(1). This includes supply fans, exhaust fans, return/relief fans, and fan-powered terminal units associated with systems providing heating or cooling capability.

TABLE C403.2.10
MINIMUM PIPE INSULATION THICKNESS (in inches)[a, c]

FLUID OPERATING TEMPERATURE RANGE AND USAGE (°F)	INSULATION CONDUCTIVITY		NOMINAL PIPE OR TUBE SIZE (inches)				
	Conductivity Btu · in./(h · ft² · °F)[b]	Mean Rating Temperature, °F	< 1	1 to < 1½	1½ to < 4	4 to < 8	≥ 8
> 350	0.32 – 0.34	250	4.5	5.0	5.0	5.0	5.0
251 – 350	0.29 – 0.32	200	3.0	4.0	4.5	4.5	4.5
201 – 250	0.27 – 0.30	150	2.5	2.5	2.5	3.0	3.0
141 – 200	0.25 – 0.29	125	1.5	1.5	2.0	2.0	2.0
105 – 140	0.21 – 0.28	100	1.0	1.0	1.5	1.5	1.5
40 – 60	0.21 – 0.27	75	0.5	0.5	1.0	1.0	1.0
< 40	0.20 – 0.26	50	0.5	1.0	1.0	1.0	1.5

For SI: 1 inch = 25.4 mm, °C = [(°F) - 32]/1.8.

a. For piping smaller than 1½ inches and located in partitions within conditioned spaces, reduction of these thicknesses by 1 inch shall be permitted (before thickness adjustment required in footnote b) but not to a thickness less than 1 inch.

b. For insulation outside the stated conductivity range, the minimum thickness (T) shall be determined as follows:

$T = r\{(1 + t/r)^{K/k} - 1\}$

where:
- T = minimum insulation thickness,
- r = actual outside radius of pipe,
- t = insulation thickness listed in the table for applicable fluid temperature and pipe size,
- K = conductivity of alternate material at mean rating temperature indicated for the applicable fluid temperature (Btu • in/h • ft² • °F) and
- k = the upper value of the conductivity range listed in the table for the applicable fluid temperature.

c. For direct-buried heating and hot water system piping, reduction of these thicknesses by 1½ inches (38 mm) shall be permitted (before thickness adjustment required in footnote b but not to thicknesses less than 1 inch (25 mm).

Single-*zone* variable air volume systems shall comply with the constant volume fan power limitation.

Exceptions:

1. Hospital, vivarium and laboratory systems that utilize flow control devices on exhaust or return to maintain space pressure relationships necessary for occupant health and safety or environmental control shall be permitted to use variable volume fan power limitation.

2. Individual exhaust fans with motor nameplate horsepower of 1 hp (0.746 kW) or less are exempt from the allowable fan horsepower requirement.

C403.2.12.2 Motor nameplate horsepower. For each fan, the fan brake horsepower shall be indicated on the construction documents and the selected motor shall be not larger than the first available motor size greater than the following:

1. For fans less than 6 bhp (4413 W), 1.5 times the fan brake horsepower.

2. For fans 6 bhp (4413 W) and larger, 1.3 times the fan brake horsepower.

3. Systems complying with Section C403.2.12.1 *fan system motor nameplate hp* (Option 1).

C403.2.12.3 Fan efficiency. Fans shall have a fan efficiency grade (FEG) of not less than 67 when determined in accordance with AMCA 205 by an *approved*, independent testing laboratory and labeled by the manufacturer. The total efficiency of the fan at the design point of operation shall be within 15 percentage points of the maximum total efficiency of the fan.

Exception: The following fans are not required to have a fan efficiency grade:

1. Fans of 5 hp (3.7 kW) or less as follows:

    1.1. Single fan with a motor nameplate horsepower of 5 hp (3.7 kW) or less, unless Exception 1.2 applies.

    1.2. Multiple fans in series or parallel that have a combined motor nameplate horsepower of 5 hp (3.7 kW) or less and are operated as the functional equivalent of a single fan.

2. Fans that are part of equipment covered under Section C403.2.3.

3. Fans included in an equipment package certified by an *approved agency* for air or energy performance.

4. Powered wall/roof ventilators.

5. Fans outside the scope of AMCA 205.

6. Fans that are intended to operate only during emergency conditions.

C403.2.13 Heating outside a building. Systems installed to provide heat outside a building shall be radiant systems.

Such heating systems shall be controlled by an occupancy sensing device or a timer switch, so that the system is automatically deenergized when no occupants are present.

TABLE C403.2.12.1(1)
FAN POWER LIMITATION

	LIMIT	CONSTANT VOLUME	VARIABLE VOLUME
Option 1: Fan system motor nameplate hp	Allowable nameplate motor hp	$hp \leq CFM_S \cdot 0.0011$	$hp \leq CFM_S \cdot 0.0015$
Option 2: Fan system bhp	Allowable fan system bhp	$bhp \leq CFM_S \cdot 0.00094 + A$	$bhp \leq CFM_S \cdot 0.0013 + A$

For SI: 1 bhp = 735.5 W, 1 hp = 745.5 W, 1 cfm = 0.4719 L/s.

where:

CFM_S = The maximum design supply airflow rate to conditioned spaces served by the system in cubic feet per minute.

hp = The maximum combined motor nameplate horsepower.

Bhp = The maximum combined fan brake horsepower.

A = Sum of $[PD \times CFM_D / 4131]$

where:

PD = Each applicable pressure drop adjustment from Table C403.2.12.1(2) in. w.c.

CFM_D = The design airflow through each applicable device from Table C403.2.12.1(2) in cubic feet per minute.

TABLE C403.2.12.1(2)
FAN POWER LIMITATION PRESSURE DROP ADJUSTMENT

DEVICE	ADJUSTMENT
Credits	
Fully ducted return and/or exhaust air systems	0.5 inch w.c. (2.15 in w.c. for laboratory and vivarium systems)
Return and/or exhaust airflow control devices	0.5 inch w.c.
Exhaust filters, scrubbers or other exhaust treatment	The pressure drop of device calculated at fan system design condition
Particulate filtration credit: MERV 9 thru 12	0.5 inch w.c.
Particulate filtration credit: MERV 13 thru 15	0.9 inch. w.c.
Particulate filtration credit: MERV 16 and greater and electronically enhanced filters	Pressure drop calculated at 2x clean filter pressure drop at fan system design condition.
Carbon and other gas-phase air cleaners	Clean filter pressure drop at fan system design condition.
Biosafety cabinet	Pressure drop of device at fan system design condition.
Energy recovery device, other than coil runaround loop	(2.2 × energy recovery effectiveness) – 0.5 inch w.c. for each airstream.
Coil runaround loop	0.6 inch w.c. for each airstream.
Evaporative humidifier/cooler in series with another cooling coil	Pressure drop of device at fan system design conditions.
Sound attenuation section (fans serving spaces with design background noise goals below NC35)	0.15 inch w.c.
Exhaust system serving fume hoods	0.35 inch w.c.
Laboratory and vivarium exhaust systems in high-rise buildings	0.25 inch w.c./100 feet of vertical duct exceeding 75 feet.
Deductions	
Systems without central cooling device	- 0.6 in. w.c.
Systems without central heating device	- 0.3 in. w.c.
Systems with central electric resistance heat	- 0.2 in. w.c.

For SI: 1 inch w.c. = 249 Pa, 1 inch = 25.4 mm.

w.c. = water column, NC = Noise criterion.

COMMERCIAL ENERGY EFFICIENCY

C403.2.14 Refrigeration equipment performance. Refrigeration equipment shall have an energy use in kWh/day not greater than the values of Tables C403.2.14(1) and C403.2.14(2) when tested and rated in accordance with AHRI Standard 1200. The energy use shall be verified through certification under an approved certification program or, where a certification program does not exist, the energy use shall be supported by data furnished by the equipment manufacturer.

TABLE C403.2.14(1)
MINIMUM EFFICIENCY REQUIREMENTS: COMMERCIAL REFRIGERATION

EQUIPMENT TYPE	APPLICATION	ENERGY USE LIMITS (kWh per day)[a]	TEST PROCEDURE
Refrigerator with solid doors	Holding Temperature	$0.10 \cdot V + 2.04$	AHRI 1200
Refrigerator with transparent doors		$0.12 \cdot V + 3.34$	
Freezers with solid doors		$0.40 \cdot V + 1.38$	
Freezers with transparent doors		$0.75 \cdot V + 4.10$	
Refrigerators/freezers with solid doors		the greater of $0.12 \cdot V + 3.34$ or 0.70	
Commercial refrigerators	Pulldown	$0.126 \cdot V + 3.51$	

a. V = volume of the chiller or frozen compartment as defined in AHAM-HRF-1.

TABLE C403.2.14(2)
MINIMUM EFFICIENCY REQUIREMENTS: COMMERCIAL REFRIGERATORS AND FREEZERS

EQUIPMENT TYPE				ENERGY USE LIMITS (kWh/day)[a,b]	TEST PROCEDURE
Equipment Class[c]	Family Code	Operating Mode	Rating Temperature		
VOP.RC.M	Vertical open	Remote condensing	Medium	$0.82 \cdot TDA + 4.07$	AHRI 1200
SVO.RC.M	Semivertical open	Remote condensing	Medium	$0.83 \cdot TDA + 3.18$	
HZO.RC.M	Horizontal open	Remote condensing	Medium	$0.35 \cdot TDA + 2.88$	
VOP.RC.L	Vertical open	Remote condensing	Low	$2.27 \cdot TDA + 6.85$	
HZO.RC.L	Horizontal open	Remote condensing	Low	$0.57 \cdot TDA + 6.88$	
VCT.RC.M	Vertical transparent door	Remote condensing	Medium	$0.22\ TDA + 1.95$	
VCT.RC.L	Vertical transparent door	Remote condensing	Low	$0.56 \cdot TDA + 2.61$	
SOC.RC.M	Service over counter	Remote condensing	Medium	$0.51 \cdot TDA + 0.11$	
VOP.SC.M	Vertical open	Self-contained	Medium	$1.74 \cdot TDA + 4.71$	
SVO.SC.M	Semivertical open	Self-contained	Medium	$1.73 \cdot TDA + 4.59$	
HZO.SC.M	Horizontal open	Self-contained	Medium	$0.77 \cdot TDA + 5.55$	
HZO.SC.L	Horizontal open	Self-contained	Low	$1.92 \cdot TDA + 7.08$	
VCT.SC.I	Vertical transparent door	Self-contained	Ice cream	$0.67 \cdot TDA + 3.29$	
VCS.SC.I	Vertical solid door	Self-contained	Ice cream	$0.38 \cdot V + 0.88$	
HCT.SC.I	Horizontal transparent door	Self-contained	Ice cream	$0.56 \cdot TDA + 0.43$	
SVO.RC.L	Semivertical open	Remote condensing	Low	$2.27 \cdot TDA + 6.85$	
VOP.RC.I	Vertical open	Remote condensing	Ice cream	$2.89 \cdot TDA + 8.7$	
SVO.RC.I	Semivertical open	Remote condensing	Ice cream	$2.89 \cdot TDA + 8.7$	
HZO.RC.I	Horizontal open	Remote condensing	Ice cream	$0.72 \cdot TDA + 8.74$	
VCT.RC.I	Vertical transparent door	Remote condensing	Ice cream	$0.66 \cdot TDA + 3.05$	
HCT.RC.M	Horizontal transparent door	Remote condensing	Medium	$0.16 \cdot TDA + 0.13$	

(continued)

TABLE C403.2.14(2)—continued
MINIMUM EFFICIENCY REQUIREMENTS: COMMERCIAL REFRIGERATORS AND FREEZERS

EQUIPMENT TYPE				ENERGY USE LIMITS (kWh/day)[a,b]	TEST PROCEDURE
Equipment Class[c]	Family Code	Operating Mode	Rating Temperature		
HCT.RC.L	Horizontal transparent door	Remote condensing	Low	$0.34 \cdot TDA + 0.26$	AHRI 1200
HCT.RC.I	Horizontal transparent door	Remote condensing	Ice cream	$0.4 \cdot TDA + 0.31$	
VCS.RC.M	Vertical solid door	Remote condensing	Medium	$0.11 \cdot V + 0.26$	
VCS.RC.L	Vertical solid door	Remote condensing	Low	$0.23 \cdot V + 0.54$	
VCS.RC.I	Vertical solid door	Remote condensing	Ice cream	$0.27 \cdot V + 0.63$	
HCS.RC.M	Horizontal solid door	Remote condensing	Medium	$0.11 \cdot V + 0.26$	
HCS.RC.L	Horizontal solid door	Remote condensing	Low	$0.23 \cdot V + 0.54$	
HCS.RC.I	Horizontal solid door	Remote condensing	Ice cream	$0.27 \cdot V + 0.63$	
HCS.RC.I	Horizontal solid door	Remote condensing	Ice cream	$0.27 \cdot V + 0.63$	
SOC.RC.L	Service over counter	Remote condensing	Low	$1.08 \cdot TDA + 0.22$	
SOC.RC.I	Service over counter	Remote condensing	Ice cream	$1.26 \cdot TDA + 0.26$	
VOP.SC.L	Vertical open	Self-contained	Low	$4.37 \cdot TDA + 11.82$	
VOP.SC.I	Vertical open	Self-contained	Ice cream	$5.55 \cdot TDA + 15.02$	
SVO.SC.L	Semivertical open	Self-contained	Low	$4.34 \cdot TDA + 11.51$	
SVO.SC.I	Semivertical open	Self-contained	Ice cream	$5.52 \cdot TDA + 14.63$	
HZO.SC.I	Horizontal open	Self-contained	Ice cream	$2.44 \cdot TDA + 9.0$	
SOC.SC.I	Service over counter	Self-contained	Ice cream	$1.76 \cdot TDA + 0.36$	
HCS.SC.I	Horizontal solid door	Self-contained	Ice cream	$0.38 \cdot V + 0.88$	

a. V = Volume of the case, as measured in accordance with Appendix C of AHRI 1200.
b. TDA = Total display area of the case, as measured in accordance with Appendix D of AHRI 1200.
c. Equipment class designations consist of a combination [(in sequential order separated by periods (AAA).(BB).(C))] of:
  (AAA) An equipment family code where:
      VOP = vertical open
      SVO = semivertical open
      HZO = horizontal open
      VCT = vertical transparent doors
      VCS = vertical solid doors
      HCT = horizontal transparent doors
      HCS = horizontal solid doors
      SOC = service over counter
  (BB) An operating mode code:
      RC = remote condensing
      SC = self-contained
  (C) A rating temperature code:
      M = medium temperature (38°F)
      L = low temperature (0°F)
      I = ice-cream temperature (15°F)
For example, ''VOP.RC.M'' refers to the "vertical-open, remote-condensing, medium-temperature" equipment class.

C403.2.15 Walk-in coolers, walk-in freezers, refrigerated warehouse coolers and refrigerated warehouse freezers. *Refrigerated warehouse coolers* and *refrigerated warehouse freezers* shall comply with this section. *Walk-in coolers* and *walk-in freezers* that are not either site assembled or site constructed shall comply with the following:

1. Be equipped with automatic door-closers that firmly close walk-in doors that have been closed to within 1 inch (25 mm) of full closure.

    Exception: Automatic closers are not required for doors more than 45 inches (1143 mm) in width or more than 7 feet (2134 mm) in height.

2. Doorways shall have strip doors, curtains, spring-hinged doors or other method of minimizing infiltration when doors are open.

3. *Walk-in coolers* and *refrigerated warehouse coolers* shall contain wall, ceiling, and door insulation of not less than R-25 and *walk-in freezers* and *refrigerated warehouse freezers* shall contain wall, ceiling and door insulation of not less than R-32.

    Exception: Glazed portions of doors or structural members need not be insulated.

4. *Walk-in freezers* shall contain floor insulation of not less than R-28.

5. Transparent reach-in doors for *walk-in freezers* and windows in *walk-in freezer* doors shall be of triple-pane glass, either filled with inert gas or with heat-reflective treated glass.

6. Windows and transparent reach-in doors for *walk-in coolers* doors shall be of double-pane or triple-pane, inert gas-filled, heat-reflective treated glass.

7. Evaporator fan motors that are less than 1 hp (0.746 kW) and less than 460 volts shall use electronically commutated motors, brushless direct-current motors, or 3-phase motors.

8. Condenser fan motors that are less than 1 hp (0.746 kW) shall use electronically commutated motors, permanent split capacitor-type motors or 3-phase motors.

9. Where antisweat heaters without antisweat heater controls are provided, they shall have a total door rail, glass and frame heater power draw of not more than 7.1 W/ft^2 (76 W/m^2) of door opening for *walk-in freezers* and 3.0 W/ft^2 (32 W/m^2) of door opening for *walk-in coolers*.

10. Where antisweat heater controls are provided, they shall reduce the energy use of the antisweat heater as a function of the relative humidity in the air outside the door or to the condensation on the inner glass pane.

11. Lights in *walk-in coolers*, *walk-in freezers*, *refrigerated warehouse coolers* and *refrigerated warehouse freezers* shall either use light sources with an efficacy of not less than 40 lumens per watt, including ballast losses, or shall use light sources with an efficacy of not less than 40 lumens per watt, including ballast losses, in conjunction with a device that turns off the lights within 15 minutes when the space is not occupied.

C403.2.16 Walk-in coolers and walk-in freezers. Site-assembled or site-constructed *walk-in coolers* and *walk-in freezers* shall comply with the following:

1. Automatic door closers shall be provided that fully close walk-in doors that have been closed to within 1 inch (25 mm) of full closure.

    Exception: Closers are not required for doors more than 45 inches (1143 mm) in width or more than 7 feet (2134 mm) in height.

2. Doorways shall be provided with strip doors, curtains, spring-hinged doors or other method of minimizing infiltration when the doors are open.

3. Walls shall be provided with insulation having a thermal resistance of not less than R-25, ceilings shall be provided with insulation having a thermal resistance of not less than R-25 and doors of *walk-in coolers* and *walk-in freezers* shall be provided with insulation having a thermal resistance of not less than R-32.

    Exception: Insulation is not required for glazed portions of doors or at structural members associated with the walls, ceiling or door frame.

4. The floor of *walk-in freezers* shall be provided with insulation having a thermal resistance of not less than R-28.

5. Transparent reach-in doors for and windows in opaque *walk-in freezer* doors shall be provided with triple-pane glass having the interstitial spaces filled with inert gas or provided with heat-reflective treated glass.

6. Transparent reach-in doors for and windows in opaque *walk-in cooler* doors shall be double-pane heat-reflective treated glass having the interstitial space gas filled.

7. Evaporator fan motors that are less than 1 hp (0.746 kW) and less than 460 volts shall be electronically commutated motors or 3-phase motors.

8. Condenser fan motors that are less than 1 hp (0.746 kW) in capacity shall be of the electronically commutated or permanent split capacitor-type or shall be 3-phase motors.

    Exception: Fan motors in *walk-in coolers* and *walk-in freezers* combined in a single enclosure greater than 3,000 square feet (279 m^2) in floor area are exempt.

9. Antisweat heaters that are not provided with antisweat heater controls shall have a total door rail, glass and frame heater power draw not greater than 7.1 W/ft^2 (76 W/m^2) of door opening for *walk-in freezers*, and not greater than 3.0 W/ft^2 (32 W/m^2) of door opening for *walk-in coolers*.

10. Antisweat heater controls shall be capable of reducing the energy use of the antisweat heater as a function of the relative humidity in the air outside the door or to the condensation on the inner glass pane.

11. Light sources shall have an efficacy of not less than 40 lumens per Watt, including any ballast losses, or shall be provided with a device that automatically turns off the lights within 15 minutes of when the *walk-in cooler* or *walk-in freezer* was last occupied.

C403.2.17 Refrigerated display cases. Site-assembled or site-constructed refrigerated display cases shall comply with the following:

1. Lighting and glass doors in refrigerated display cases shall be controlled by one of the following:

    1.1. Time switch controls to turn off lights during nonbusiness hours. Timed overrides for display cases shall turn the lights on for up to 1 hour and shall automatically time out to turn the lights off.

    1.2. Motion sensor controls on each display case section that reduce lighting power by at least 50 percent within 3 minutes after the area within the sensor range is vacated.

2. Low-temperature display cases shall incorporate temperature-based defrost termination control with a time-limit default. The defrost cycle shall terminate first on an upper temperature limit breach and second upon a time limit breach.

3. Antisweat heater controls shall reduce the energy use of the antisweat heater as a function of the relative humidity in the air outside the door or to the condensation on the inner glass pane.

C403.3 Economizers (Prescriptive). Each cooling system shall include either an air or water economizer complying with Sections C403.3.1 through C403.3.4.

Exceptions: Economizers are not required for the systems listed below.

1. In cooling systems for buildings located in *Climate Zones* 1A and 1B.

2. In *climate zones* other than 1A and 1B, where individual fan cooling units have a capacity of less than 54,000 Btu/h (15.8 kW) and meet one of the following:

    2.1. Have direct expansion cooling coils.

    2.2. The total chilled water system capacity less the capacity of fan units with air economizers is less than the minimum specified in Table C403.3(1).

The total supply capacity of all fan-cooling units not provided with economizers shall not exceed 20 percent of the total supply capacity of all fan-cooling units in the building or 300,000 Btu/h (88 kW), whichever is greater.

3. Where more than 25 percent of the air designed to be supplied by the system is to spaces that are designed to be humidified above 35°F (1.7°C) dewpoint temperature to satisfy process needs.

4. Systems that serve *residential* spaces where the system capacity is less than five times the requirement listed in Table C403.3(1).

5. Systems expected to operate less than 20 hours per week.

6. Where the use of *outdoor air* for cooling will affect supermarket open refrigerated casework systems.

7. Where the cooling *efficiency* meets or exceeds the *efficiency* requirements in Table C403.3(2).

8. Chilled-water cooling systems that are passive (without a fan) or use induction where the total chilled water system capacity less the capacity of fan units with air economizers is less than the minimum specified in Table C403.3(1).

9. Systems that include a heat recovery system in accordance with Section C403.4.5.

TABLE C403.3(2)
EQUIPMENT EFFICIENCY PERFORMANCE EXCEPTION FOR ECONOMIZERS

CLIMATE ZONES	COOLING EQUIPMENT PERFORMANCE IMPROVEMENT (EER OR IPLV)
2B	10% efficiency improvement
3B	15% efficiency improvement
4B	20% efficiency improvement

C403.3.1 Integrated economizer control. Economizer systems shall be integrated with the mechanical cooling system and be capable of providing partial cooling even where additional mechanical cooling is required to provide the remainder of the cooling load. Controls shall not be capable of creating a false load in the mechanical cooling systems by limiting or disabling the economizer or any other means, such as hot gas bypass, except at the lowest stage of mechanical cooling.

TABLE C403.3(1)
MINIMUM CHILLED-WATER SYSTEM COOLING CAPACITY FOR DETERMINING ECONOMIZER COOLING REQUIREMENTS

CLIMATE ZONES (COOLING)	TOTAL CHILLED-WATER SYSTEM CAPACITY LESS CAPACITY OF COOLING UNITS WITH AIR ECONOMIZERS	
	Local Water-cooled Chilled-water Systems	Air-cooled Chilled-water Systems or District Chilled-Water Systems
1a	No economizer requirement	No economizer requirement
1b, 2a, 2b	960,000 Btu/h	1,250,000 Btu/h
3a, 3b, 3c, 4a, 4b, 4c	720,000 Btu/h	940,000 Btu/h
5a, 5b, 5c, 6a, 6b, 7, 8	1,320,000 Btu/h	1,720,000 Btu/h

For SI:1 British thermal unit per hour = 0.2931 W.

Units that include an air economizer shall comply with the following:

1. Unit controls shall have the mechanical cooling capacity control interlocked with the air economizer controls such that the outdoor air damper is at the 100-percent open position when mechanical cooling is on and the outdoor air damper does not begin to close to prevent coil freezing due to minimum compressor run time until the leaving air temperature is less than 45°F (7°C).
2. Direct expansion (DX) units that control 75,000 Btu/h (22 kW) or greater of rated capacity of the capacity of the mechanical cooling directly based on occupied space temperature shall have not fewer than two stages of mechanical cooling capacity
3. Other DX units, including those that control space temperature by modulating the airflow to the space, shall be in accordance with Table C403.3.1.

C403.3.2 Economizer heating system impact. HVAC system design and economizer controls shall be such that economizer operation does not increase building heating energy use during normal operation.

Exception: Economizers on variable air volume (VAV) systems that cause *zone* level heating to increase due to a reduction in supply air temperature.

C403.3.3 Air economizers. Air economizers shall comply with Sections C403.3.3.1 through C403.3.3.5.

C403.3.3.1 Design capacity. Air economizer systems shall be capable of modulating *outdoor air* and return air dampers to provide up to 100 percent of the design supply air quantity as *outdoor air* for cooling.

C403.3.3.2 Control signal. Economizer dampers shall be capable of being sequenced with the mechanical cooling equipment and shall not be controlled by only mixed-air temperature.

Exception: The use of mixed-air temperature limit control shall be permitted for systems controlled from space temperature (such as single-*zone* systems).

C403.3.3.3 High-limit shutoff. Air economizers shall be capable of automatically reducing *outdoor air* intake to the design minimum *outdoor air* quantity when *outdoor air* intake will no longer reduce cooling energy usage. High-limit shutoff control types for specific climates shall be chosen from Table C403.3.3.3. High-limit shutoff control settings for these control types shall be those specified in Table C403.3.3.3.

C403.3.3.4 Relief of excess outdoor air. Systems shall be capable of relieving excess *outdoor air* during air economizer operation to prevent overpressurizing the building. The relief air outlet shall be located to avoid recirculation into the building.

TABLE C403.3.1
DX COOLING STAGE REQUIREMENTS FOR MODULATING AIRFLOW UNITS

RATING CAPACITY	MINIMUM NUMBER OF MECHANICAL COOLING STAGES	MINIMUM COMPRESSOR DISPLACEMENT[a]
≥ 65,000 Btu/h and < 240,000 Btu/h	3 stages	≤ 35% of full load
≥ 240,000 Btu/h	4 stages	≤ 25% full load

For SI: 1 British thermal unit per hour = 0.2931 W.

a. For *mechanical cooling* stage control that does not use variable compressor displacement, the percent displacement shall be equivalent to the mechanical cooling capacity reduction evaluated at the full load rating conditions for the compressor.

TABLE C403.3.3.3
HIGH-LIMIT SHUTOFF CONTROL SETTING FOR AIR ECONOMIZERS[b]

DEVICE TYPE	CLIMATE ZONE	REQUIRED HIGH LIMIT (ECONOMIZER OFF WHEN):	
		Equation	Description
Fixed dry bulb	1B, 2B, 3B, 3C, 4B, 4C, 5B, 5C, 6B, 7, 8	$T_{OA} > 75°F$	Outdoor air temperature exceeds 75°F
	5A, 6A	$T_{OA} > 70°F$	Outdoor air temperature exceeds 70°F
	1A, 2A, 3A, 4A	$T_{OA} > 65°F$	Outdoor air temperature exceeds 65°F
Differential dry bulb	1B, 2B, 3B, 3C, 4B, 4C, 5A, 5B, 5C, 6A, 6B, 7, 8	$T_{OA} > T_{RA}$	Outdoor air temperature exceeds return air temperature
Fixed enthalpy with fixed dry-bulb temperatures	All	$h_{OA} > 28$ Btu/lb[a] or $T_{OA} > 75°F$	Outdoor air enthalpy exceeds 28 Btu/lb of dry air[a] or Outdoor air temperature exceeds 75°F
Differential enthalpy with fixed dry-bulb temperature	All	$h_{OA} > h_{RA}$ or $T_{OA} > 75°F$	Outdoor air enthalpy exceeds return air enthalpy or Outdoor air temperature exceeds 75°F

For SI: 1 foot = 305 mm, °C = (°F - 32)/1.8, 1 Btu/lb = 2.33 kJ/kg.

a. At altitudes substantially different than sea level, the fixed enthalpy limit shall be set to the enthalpy value at 75°F and 50-percent relative humidity. As an example, at approximately 6,000 feet elevation, the fixed enthalpy limit is approximately 30.7 Btu/lb.

b. Devices with selectable setpoints shall be capable of being set to within 2°F and 2 Btu/lb of the setpoint listed.

C403.3.3.5 Economizer dampers. Return, exhaust/relief and outdoor air dampers used in economizers shall comply with Section C403.2.4.3.

C403.3.4 Water-side economizers. Water-side economizers shall comply with Sections C403.3.4.1 and C403.3.4.2.

C403.3.4.1 Design capacity. Water economizer systems shall be capable of cooling supply air by indirect evaporation and providing up to 100 percent of the expected system cooling load at *outdoor air* temperatures of not greater than 50°F (10°C) dry bulb/45°F (7°C) wet bulb.

Exceptions:

1. Systems primarily serving computer rooms in which 100 percent of the expected system cooling load at 40°F (4°C) dry bulb/35°F (1.7°C) wet bulb is met with evaporative water economizers.

2. Systems primarily serving computer rooms with dry cooler water economizers which satisfy 100 percent of the expected system cooling load at 35°F (1.7°C) dry bulb.

3. Systems where dehumidification requirements cannot be met using outdoor air temperatures of 50°F (10°C) dry bulb/45°F (7°C) wet bulb and where 100 percent of the expected system cooling load at 45°F (7°C) dry bulb/40°F (4°C) wet bulb is met with evaporative water economizers.

C403.3.4.2 Maximum pressure drop. Precooling coils and water-to-water heat exchangers used as part of a water economizer system shall either have a water-side pressure drop of less than 15 feet (45 kPa) of water or a secondary loop shall be created so that the coil or heat exchanger pressure drop is not seen by the circulating pumps when the system is in the normal cooling (noneconomizer) mode.

C403.4 Hydronic and multiple-zone HVAC systems controls and equipment. (Prescriptive). Hydronic and multiple-zone HVAC system controls and equipment shall comply with this section.

C403.4.1 Fan control. Controls shall be provided for fans in accordance with Sections C403.4.1.1 through C403.4.1.3.

C403.4.1.1 Fan airflow control. Each cooling system listed in Table C403.4.1.1 shall be designed to vary the indoor fan airflow as a function of load and shall comply with the following requirements:

1. Direct expansion (DX) and chilled water cooling units that control the capacity of the mechanical cooling directly based on space temperature shall have not fewer than two stages of fan control. Low or minimum speed shall not be greater than 66 percent of full speed. At low or minimum speed, the fan system shall draw not more than 40 percent of the fan power at full fan speed. Low or minimum speed shall be used during periods of low cooling load and ventilation-only operation.

2. Other units including DX cooling units and chilled water units that control the space temperature by modulating the airflow to the space shall have modulating fan control. Minimum speed shall be not greater than 50 percent of full speed. At minimum speed the fan system shall draw not more than 30 percent of the power at full fan speed. Low or minimum speed shall be used during periods of low cooling load and ventilation-only operation.

3. Units that include an airside economizer in accordance with Section C403.3 shall have not fewer than two speeds of fan control during economizer operation

Exceptions:

1. Modulating fan control is not required for chilled water and evaporative cooling units with fan motors of less than 1 hp (0.746 kW) where the units are not used to provide *ventilation* air and the indoor fan cycles with the load.

2. Where the volume of outdoor air required to comply with the *ventilation* requirements of the *International Mechanical Code* at low speed exceeds the air that would be delivered at the speed defined in Section C403.4.1, the minimum speed shall be selected to provide the required *ventilation air*.

TABLE C403.4.1.1
EFFECTIVE DATES FOR FAN CONTROL

COOLING SYSTEM TYPE	FAN MOTOR SIZE	MECHANICAL COOLING CAPACITY
DX cooling	Any	≥ 75,000 Btu/h (before 1/1/2016)
		≥ 65,000 Btu/h (after 1/1/2016)
Chilled water and evaporative cooling	≥ 5 hp	Any
	≥ $^1/_4$ hp	Any

For SI: 1 British thermal unit per hour = 0.2931 W; 1 hp = 0.746 kW.

C403.4.1.2 Static pressure sensor location. Static pressure sensors used to control VAV fans shall be located such that the controller set point is not greater than 1.2 inches w.c. (299 Pa). Where this results in one or more sensors being located downstream of major duct splits, not less than one sensor shall be located on each major branch to ensure that static pressure can be maintained in each branch.

C403.4.1.3 Set points for direct digital control. For systems with direct digital control of individual zones reporting to the central control panel, the static pressure set point shall be reset based on the *zone* requiring the most pressure. In such case, the set point is reset lower until one *zone* damper is nearly wide open. The direct digital controls shall be capable of monitoring *zone* damper positions or shall have an alternative method of indicating the need for static pressure that is capable of all of the following:

1. Automatically detecting any *zone* that excessively drives the reset logic.
2. Generating an alarm to the system operational location.
3. Allowing an operator to readily remove one or more *zones* from the reset algorithm.

C403.4.2 Hydronic systems controls. The heating of fluids that have been previously mechanically cooled and the cooling of fluids that have been previously mechanically heated shall be limited in accordance with Sections C403.4.2.1 through C403.4.2.3. Hydronic heating systems comprised of multiple-packaged boilers and designed to deliver conditioned water or steam into a common distribution system shall include automatic controls capable of sequencing operation of the boilers. Hydronic heating systems comprised of a single boiler and greater than 500,000 Btu/h (146.5 kW) input design capacity shall include either a multistaged or modulating burner.

C403.4.2.1 Three-pipe system. Hydronic systems that use a common return system for both hot water and chilled water are prohibited.

C403.4.2.2 Two-pipe changeover system. Systems that use a common distribution system to supply both heated and chilled water shall be designed to allow a dead band between changeover from one mode to the other of not less than 15°F (8.3°C) outside air temperatures; be designed to and provided with controls that will allow operation in one mode for not less than 4 hours before changing over to the other mode; and be provided with controls that allow heating and cooling supply temperatures at the changeover point to be not more than 30°F (16.7°C) apart.

C403.4.2.3 Hydronic (water loop) heat pump systems. Hydronic heat pump systems shall comply with Sections C403.4.2.3.1 through C403.4.2.3.2.

C403.4.2.3.1 Temperature dead band. Hydronic heat pumps connected to a common heat pump water loop with central devices for heat rejection and heat addition shall have controls that are capable of providing a heat pump water supply temperature dead band of not less than 20°F (11°C) between initiation of heat rejection and heat addition by the central devices.

> **Exception:** Where a system loop temperature optimization controller is installed and can determine the most efficient operating temperature based on realtime conditions of demand and capacity, dead bands of less than 20°F (11°C) shall be permitted.

C403.4.2.3.2 Heat rejection. Heat rejection equipment shall comply with Sections C403.4.2.3.2.1 and C403.4.2.3.2.2.

> **Exception:** Where it can be demonstrated that a heat pump system will be required to reject heat throughout the year.

C403.4.2.3.2.1 Climate zones 3 and 4. For *Climate Zones* 3 and 4:

1. Where a closed-circuit cooling tower is used directly in the heat pump loop, either an automatic valve shall be installed to bypass all but a minimal flow of water around the tower, or lower leakage positive closure dampers shall be provided.
2. Where an open-circuit tower is used directly in the heat pump loop, an automatic valve shall be installed to bypass all heat pump water flow around the tower.
3. Where an open- or closed-circuit cooling tower is used in conjunction with a separate heat exchanger to isolate the cooling tower from the heat pump loop, then heat loss shall be controlled by shutting down the circulation pump on the cooling tower loop.

C403.4.2.3.2.2 Climate zones 5 through 8. For *Climate Zones* 5 through 8, where an open- or closed-circuit cooling tower is used, a separate heat exchanger shall be provided to isolate the cooling tower from the heat pump loop, and heat loss shall be controlled by shutting down the circulation pump on the cooling tower loop and providing an automatic valve to stop the flow of fluid.

C403.4.2.3.3 Two-position valve. Each hydronic heat pump on the hydronic system having a total pump system power exceeding 10 hp (7.5 kW) shall have a two-position valve.

C403.4.2.4 Part-load controls. Hydronic systems greater than or equal to 500,000 Btu/h (146.5 kW) in design output capacity supplying heated or chilled water to comfort conditioning systems shall include controls that have the capability to do all of the following:

1. Automatically reset the supply-water temperatures in response to varying building heating and cooling demand using coil valve position, zone-return water temperature, building-return water temperature or outside air temperature. The temperature shall be capable of being reset by not less than 25 percent of the design supply-to-return water temperature difference.
2. Automatically vary fluid flow for hydronic systems with a combined motor capacity of 10 hp (7.5 kW) or larger with three or more control valves or other devices by reducing the system design flow rate by not less than 50 percent by

designed valves that modulate or step open and close, or pumps that modulate or turn on and off as a function of load.

3. Automatically vary pump flow on chilled-water systems and heat rejection loops serving water-cooled unitary air conditioners with a combined motor capacity of 10 hp (7.5 kW) or larger by reducing pump design flow by not less than 50 percent, utilizing adjustable speed drives on pumps, or multiple-staged pumps where not less than one-half of the total pump horsepower is capable of being automatically turned off. Pump flow shall be controlled to maintain one control valve nearly wide open or to satisfy the minimum differential pressure.

Exceptions:

1. Supply-water temperature reset for chilled-water systems supplied by off-site district chilled water or chilled water from ice storage systems.

2. Minimum flow rates other than 50 percent as required by the equipment manufacturer for proper operation of equipment where using flow bypass or end-of-line 3-way valves.

3. Variable pump flow on dedicated equipment circulation pumps where configured in primary/secondary design to provide the minimum flow requirements of the equipment manufacturer for proper operation of equipment.

C403.4.2.5 Boiler turndown. *Boiler systems* with design input of greater than 1,000,000 Btu/h (293 kW) shall comply with the turndown ratio specified in Table C403.4.2.5.

The system turndown requirement shall be met through the use of multiple single input boilers, one or more *modulating boilers* or a combination of single input and *modulating boilers*.

C403.4.2.6 Pump isolation. Chilled water plants including more than one chiller shall have the capability to reduce flow automatically through the chiller plant when a chiller is shut down. Chillers piped in series for the purpose of increased temperature differential shall be considered as one chiller.

Boiler plants including more than one boiler shall have the capability to reduce flow automatically through the boiler plant when a boiler is shut down.

TABLE C403.4.2.5
BOILER TURNDOWN

BOILER SYSTEM DESIGN INPUT (Btu/h)	MINIMUM TURNDOWN RATIO
≥ 1,000,000 and less than or equal to 5,000,000	3 to 1
> 5,000,000 and less than or equal to 10,000,000	4 to 1
> 10,000,000	5 to 1

For SI: 1 British thermal unit per hour = 0.2931 W.

C403.4.3 Heat rejection equipment. Each fan powered by a motor of 7.5 hp (5.6 kW) or larger shall have the capability to operate that fan at two-thirds of full speed or less, and shall have controls that automatically change the fan speed to control the leaving fluid temperature or condensing temperature/pressure of the heat rejection device.

Exception: Factory-installed heat rejection devices within HVAC equipment tested and rated in accordance with Tables C403.2.3(6) and C403.2.3(7).

C403.4.3.1 General. Heat rejection equipment such as air-cooled condensers, dry coolers, open-circuit cooling towers, closed-circuit cooling towers and evaporative condensers used for comfort cooling applications shall comply with this section.

Exception: Heat rejection devices where energy usage is included in the equipment efficiency ratings listed in Tables C403.2.3(6) and C403.2.3(7).

C403.4.3.2 Fan speed control. The fan speed shall be controlled as provided in Sections C403.4.3.2.1 and C403.4.3.2.2.

C403.4.3.2.1 Fan motors not less than 7.5 hp. Each fan powered by a motor of 7.5 hp (5.6 kW) or larger shall have the capability to operate that fan at two-thirds of full speed or less, and shall have controls that automatically change the fan speed to control the leaving fluid temperature or condensing temperature/pressure of the heat rejection device.

Exception: The following fan motors over 7.5 hp (5.6 kW) are exempt:

1. Condenser fans serving multiple refrigerant circuits.

2. Condenser fans serving flooded condensers.

3. Installations located in *Climate Zones* 1 and 2.

C403.4.3.2.2 Multiple-cell heat rejection equipment. Multiple-cell heat rejection equipment with variable speed fan drives shall be controlled in both of the following manners:

1. To operate the maximum number of fans allowed that comply with the manufacturer's requirements for all system components.

2. So all fans can operate at the same fan speed required for the instantaneous cooling duty, as opposed to staged (on/off) operation.

Minimum fan speed shall be the minimum allowable speed of the fan drive system in accordance with the manufacturer's recommendations.

C403.4.3.3 Limitation on centrifugal fan open-circuit cooling towers. Centrifugal fan open-circuit cooling towers with a combined rated capacity of 1,100 gpm (4164 L/m) or greater at 95°F (35°C) condenser water return, 85°F (29°C) condenser water supply, and 75°F (24°C) outdoor air wet-bulb temperature shall meet the energy efficiency requirement for axial fan open-circuit cooling towers listed in Table C403.2.3(8).

Exception: Centrifugal open-circuit cooling towers that are designed with inlet or discharge ducts or require external sound attenuation.

C403.4.3.4 Tower flow turndown. Open-circuit cooling towers used on water-cooled chiller systems that are configured with multiple- or variable-speed condenser water pumps shall be designed so that all open-circuit cooling tower cells can be run in parallel with the larger of the flow that is produced by the smallest pump at its minimum expected flow rate or at 50 percent of the design flow for the cell.

C403.4.4 Requirements for complex mechanical systems serving multiple zones. Sections C403.4.4.1 through C403.4.6.4 shall apply to complex mechanical systems serving multiple zones. Supply air systems serving multiple zones shall be variable air volume (VAV) systems that, during periods of occupancy, are designed and capable of being controlled to reduce primary air supply to each *zone* to one of the following before reheating, recooling or mixing takes place:

1. Thirty percent of the maximum supply air to each *zone*.
2. Three hundred cfm (142 L/s) or less where the maximum flow rate is less than 10 percent of the total fan system supply airflow rate.
3. The minimum ventilation requirements of Chapter 4 of the *International Mechanical Code*.
4. Any higher rate that can be demonstrated to reduce overall system annual energy use by offsetting reheat/recool energy losses through a reduction in *outdoor air* intake for the system, as *approved* by the *code official*.
5. The airflow rate required to comply with applicable codes or accreditation standards, such as pressure relationships or minimum air change rates.

Exception: The following individual *zones* or entire air distribution systems are exempted from the requirement for VAV control:

1. *Zones* or supply air systems where not less than 75 percent of the energy for reheating or for providing warm air in mixing systems is provided from a site-recovered or site-solar energy source.
2. *Zones* where special humidity levels are required to satisfy process needs.
3. *Zones* with a peak supply air quantity of 300 cfm (142 L/s) or less and where the flow rate is less than 10 percent of the total fan system supply airflow rate.
4. *Zones* where the volume of air to be reheated, recooled or mixed is not greater than the volume of outside air required to provide the minimum ventilation requirements of Chapter 4 of the *International Mechanical Code*.
5. *Zones* or supply air systems with thermostatic and humidistatic controls capable of operating in sequence the supply of heating and cooling energy to the *zones* and which are capable of preventing reheating, recooling, mixing or simultaneous supply of air that has been previously cooled, either mechanically or through the use of economizer systems, and air that has been previously mechanically heated.

C403.4.4.1 Single-duct VAV systems, terminal devices. Single-duct VAV systems shall use terminal devices capable of reducing the supply of primary supply air before reheating or recooling takes place.

C403.4.4.2 Dual-duct and mixing VAV systems, terminal devices. Systems that have one warm air duct and one cool air duct shall use terminal devices that are capable of reducing the flow from one duct to a minimum before mixing of air from the other duct takes place.

C403.4.4.3 Single-fan dual-duct and mixing VAV systems, economizers. Individual dual-duct or mixing heating and cooling systems with a single fan and with total capacities greater than 90,000 Btu/h [(26.4 kW) 7.5 tons] shall not be equipped with air economizers.

C403.4.4.4 Fractional hp fan motors. Motors for fans that are not less than $1/12$ hp (0.082 kW) and less than 1 hp (0.746 kW) shall be electronically commutated motors or shall have a minimum motor efficiency of 70 percent, rated in accordance with DOE 10 CFR 431. These motors shall also have the means to adjust motor speed for either balancing or remote control. The use of belt-driven fans to sheave adjustments for airflow balancing instead of a varying motor speed shall be permitted.

Exceptions: The following motors are not required to comply with this section:

1. Motors in the airstream within fan coils and terminal units that only provide heating to the space served.
2. Motors in space-conditioning equipment that comply with Section 403.2.3 or C403.2.12.
3. Motors that comply with Section C405.8.

C403.4.4.5 Supply-air temperature reset controls. Multiple-*zone* HVAC systems shall include controls that automatically reset the supply-air temperature in response to representative building loads, or to outdoor air temperature. The controls shall be capable of resetting the supply air temperature not less than 25 percent of the difference between the design supply-air temperature and the design room air temperature.

Exceptions:

1. Systems that prevent reheating, recooling or mixing of heated and cooled supply air.
2. Seventy-five percent of the energy for reheating is from site-recovered or site-solar energy sources.
3. *Zones* with peak supply air quantities of 300 cfm (142 L/s) or less.

C403.4.4.6 Multiple-zone VAV system ventilation optimization control. Multiple-zone VAV systems with direct digital control of individual zone boxes reporting to a central control panel shall have automatic controls configured to reduce outdoor air intake flow below design rates in response to changes in system *ventilation* efficiency (E_v) as defined by the *International Mechanical Code*.

Exceptions:

1. VAV systems with zonal transfer fans that recirculate air from other zones without directly mixing it with outdoor air, dual-duct dual-fan VAV systems, and VAV systems with fan-powered terminal units.

2. Systems having exhaust air energy recovery complying with Section C403.2.7.

3. Systems where total design exhaust airflow is more than 70 percent of total design outdoor air intake flow requirements.

C403.4.5 Heat recovery for service water heating. Condenser heat recovery shall be installed for heating or reheating of service hot water provided that the facility operates 24 hours a day, the total installed heat capacity of water-cooled systems exceeds 6,000,000 Btu/hr (1 758 kW) of heat rejection, and the design service water heating load exceeds 1,000,000 Btu/h (293 kW).

The required heat recovery system shall have the capacity to provide the smaller of the following:

1. Sixty percent of the peak heat rejection load at design conditions.

2. The preheating required to raise the peak service hot water draw to 85°F (29°C).

Exceptions:

1. Facilities that employ condenser heat recovery for space heating or reheat purposes with a heat recovery design exceeding 30 percent of the peak water-cooled condenser load at design conditions.

2. Facilities that provide 60 percent of their service water heating from site solar or site recovered energy or from other sources.

C403.4.6 Hot gas bypass limitation. Cooling systems shall not use hot gas bypass or other evaporator pressure control systems unless the system is designed with multiple steps of unloading or continuous capacity modulation. The capacity of the hot gas bypass shall be limited as indicated in Table C403.4.6, as limited by Section C403.3.1.

TABLE C403.4.6
MAXIMUM HOT GAS BYPASS CAPACITY

RATED CAPACITY	MAXIMUM HOT GAS BYPASS CAPACITY (% of total capacity)
≤ 240,000 Btu/h	50
> 240,000 Btu/h	25

For SI: 1 British thermal unit per hour = 0.2931 W.

C403.5 Refrigeration systems. Refrigerated display cases, *walk-in coolers* or *walk-in freezers* that are served by remote compressors and remote condensers not located in a *condensing unit*, shall comply with Sections C403.5.1 and C403.5.2.

Exception: Systems where the working fluid in the refrigeration cycle goes through both subcritical and supercritical states (transcritical) or that use ammonia refrigerant are exempt.

C403.5.1 Condensers serving refrigeration systems. Fan-powered condensers shall comply with the following:

1. The design *saturated condensing temperatures* for air-cooled condensers shall not exceed the design dry-bulb temperature plus 10°F (5.6°C) for low-*temperature refrigeration systems*, and the design dry-bulb temperature plus 15°F (8°C) for *medium temperature refrigeration systems* where the *saturated condensing temperature* for blend refrigerants shall be determined using the average of liquid and vapor temperatures as converted from the condenser drain pressure.

2. Condenser fan motors that are less than 1 hp (0.75 kW) shall use electronically commutated motors, permanent split-capacitor-type motors or 3-phase motors.

3. Condenser fans for air-cooled condensers, evaporatively cooled condensers, air- or water-cooled fluid coolers or cooling towers shall reduce fan motor demand to not more than 30 percent of design wattage at 50 percent of design air volume, and incorporate one of the following continuous variable speed fan control approaches:

    3.1. Refrigeration system condenser control for air-cooled condensers shall use variable setpoint control logic to reset the condensing temperature setpoint in response to ambient dry-bulb temperature.

    3.2. Refrigeration system condenser control for evaporatively cooled condensers shall use variable setpoint control logic to reset the condensing temperature setpoint in response to ambient wet-bulb temperature.

4. Multiple fan condensers shall be controlled in unison.

5. The minimum condensing temperature setpoint shall be not greater than 70°F (21°C).

C403.5.2 Compressor systems. Refrigeration compressor systems shall comply with the following:

1. Compressors and multiple-compressor system suction groups shall include control systems that use floating suction pressure control logic to reset the target suction pressure temperature based on the temperature requirements of the attached refrigeration display cases or walk-ins.

    Exception: Controls are not required for the following:

    1. Single-compressor systems that do not have variable capacity capability.

    2. Suction groups that have a design saturated suction temperature of 30°F (-1.1°C) or higher, suction groups that comprise the high stage of a two-stage or cascade system, or suction groups that primarily serve chillers for secondary cooling fluids.

2. Liquid subcooling shall be provided for all low-temperature compressor systems with a design cooling capacity equal to or greater than 100,000 Btu/hr (29.3 kW) with a design-saturated suction temperature of -10°F (-23°C) or lower. The sub-cooled liquid temperature shall be controlled at a maximum temperature setpoint of 50°F (10°C) at the exit of the subcooler using either compressor economizer (interstage) ports or a separate compressor suction group operating at a saturated suction temperature of 18°F (-7.8°C) or higher.

    2.1. Insulation for liquid lines with a fluid operating temperature less than 60°F (15.6°C) shall comply with Table C403.2.10.

3. Compressors that incorporate internal or external crankcase heaters shall provide a means to cycle the heaters off during compressor operation.

SECTION C404
SERVICE WATER HEATING (MANDATORY)

C404.1 General. This section covers the minimum efficiency of, and controls for, service water-heating equipment and insulation of service hot water piping.

C404.2 Service water-heating equipment performance efficiency. Water-heating equipment and hot water storage tanks shall meet the requirements of Table C404.2. The efficiency shall be verified through data furnished by the manufacturer of the equipment or through certification under an *approved* certification program. Water-heating equipment also intended to be used to provide space heating shall meet the applicable provisions of Table C404.2.

C404.2.1 High input-rated service water-heating systems. Gas-fired water-heating equipment installed in new buildings shall be in compliance with this section. Where a singular piece of water-heating equipment serves the entire building and the input rating of the equipment is 1,000,000 Btu/h (293 kW) or greater, such equipment shall have a thermal efficiency, E_t, of not less than 90 percent. Where multiple pieces of water-heating equipment serve the building and the combined input rating of the water-heating equipment is 1,000,000 Btu/h (293 kW) or greater, the combined input-capacity-weighted-average thermal efficiency, E_t, shall be not less than 90 percent.

Exceptions:

1. Where 25 percent of the annual *service water-heating* requirement is provided by site-solar or site-recovered energy, the minimum thermal efficiency requirements of this section shall not apply.

2. The input rating of water heaters installed in individual dwelling units shall not be required to be included in the total input rating of *service water-heating* equipment for a building.

3. The input rating of water heaters with an input rating of not greater than 100,000 Btu/h (29.3 kW) shall not be required to be included in the total input rating of *service water-heating* equipment for a building.

C404.3 Heat traps. Water-heating equipment not supplied with integral heat traps and serving noncirculating systems shall be provided with heat traps on the supply and discharge piping associated with the equipment.

C404.4 Insulation of piping. Piping from a water heater to the termination of the heated water fixture supply pipe shall be insulated in accordance with Table C403.2.10. On both the inlet and outlet piping of a storage water heater or heated water storage tank, the piping to a heat trap or the first 8 feet (2438 mm) of piping, whichever is less, shall be insulated. Piping that is heat traced shall be insulated in accordance with Table C403.2.10 or the heat trace manufacturer's instructions. Tubular pipe insulation shall be installed in accordance with the insulation manufacturer's instructions. Pipe insulation shall be continuous except where the piping passes through a framing member. The minimum insulation thickness requirements of this section shall not supersede any greater insulation thickness requirements necessary for the protection of piping from freezing temperatures or the protection of personnel against external surface temperatures on the insulation.

Exception: Tubular pipe insulation shall not be required on the following:

1. The tubing from the connection at the termination of the fixture supply piping to a plumbing fixture or plumbing appliance.

2. Valves, pumps, strainers and threaded unions in piping that is 1 inch (25 mm) or less in nominal diameter.

3. Piping from user-controlled shower and bath mixing valves to the water outlets.

4. Cold-water piping of a demand recirculation water system.

5. Tubing from a hot drinking-water heating unit to the water outlet.

6. Piping at locations where a vertical support of the piping is installed.

7. Piping surrounded by building insulation with a thermal resistance (*R*-value) of not less than R-3.

TABLE C404.2
MINIMUM PERFORMANCE OF WATER-HEATING EQUIPMENT

EQUIPMENT TYPE	SIZE CATEGORY (input)	SUBCATEGORY OR RATING CONDITION	PERFORMANCE REQUIRED[a, b]	TEST PROCEDURE
Water heaters, electric	≤ 12 kW[d]	Resistance	$0.97 - 0.00132V$, EF	DOE 10 CFR Part 430
	> 12 kW	Resistance	$(0.3 + 27/V_m)$, %/h	ANSI Z21.10.3
	≤ 24 amps and ≤ 250 volts	Heat pump	$0.93 - 0.00132V$, EF	DOE 10 CFR Part 430
Storage water heaters, gas	≤ 75,000 Btu/h	≥ 20 gal	$0.67 - 0.0019V$, EF	DOE 10 CFR Part 430
	> 75,000 Btu/h and ≤ 155,000 Btu/h	< 4,000 Btu/h/gal	$80\% E_t$ $(Q/800 + 110\sqrt{V})$SL, Btu/h	ANSI Z21.10.3
	> 155,000 Btu/h	< 4,000 Btu/h/gal	$80\% E_t$ $(Q/800 + 110\sqrt{V})$SL, Btu/h	
Instantaneous water heaters, gas	> 50,000 Btu/h and < 200,000 Btu/h[c]	≥ 4,000 (Btu/h)/gal and < 2 gal	$0.62 - 0.0019V$, EF	DOE 10 CFR Part 430
	≥ 200,000 Btu/h	≥ 4,000 Btu/h/gal and < 10 gal	$80\% E_t$	ANSI Z21.10.3
	≥ 200,000 Btu/h	≥ 4,000 Btu/h/gal and ≥ 10 gal	$80\% E_t$ $(Q/800 + 110\sqrt{V})$SL, Btu/h	
Storage water heaters, oil	≤ 105,000 Btu/h	≥ 20 gal	$0.59 - 0.0019V$, EF	DOE 10 CFR Part 430
	≥ 105,000 Btu/h	< 4,000 Btu/h/gal	$80\% E_t$ $(Q/800 + 110\sqrt{V})$SL, Btu/h	ANSI Z21.10.3
Instantaneous water heaters, oil	≤ 210,000 Btu/h	≥ 4,000 Btu/h/gal and < 2 gal	$0.59 - 0.0019V$, EF	DOE 10 CFR Part 430
	> 210,000 Btu/h	≥ 4,000 Btu/h/gal and < 10 gal	$80\% E_t$	ANSI Z21.10.3
	> 210,000 Btu/h	≥ 4,000 Btu/h/gal and ≥ 10 gal	$78\% E_t$ $(Q/800 + 110\sqrt{V})$SL, Btu/h	
Hot water supply boilers, gas and oil	≥ 300,000 Btu/h and < 12,500,000 Btu/h	≥ 4,000 Btu/h/gal and < 10 gal	$80\% E_t$	
Hot water supply boilers, gas	≥ 300,000 Btu/h and < 12,500,000 Btu/h	≥ 4,000 Btu/h/gal and ≥ 10 gal	$80\% E_t$ $(Q/800 + 110\sqrt{V})$SL, Btu/h	ANSI Z21.10.3
Hot water supply boilers, oil	> 300,000 Btu/h and < 12,500,000 Btu/h	> 4,000 Btu/h/gal and > 10 gal	$78\% E_t$ $(Q/800 + 110\sqrt{V})$SL, Btu/h	
Pool heaters, gas and oil	All	—	$82\% E_t$	ASHRAE 146
Heat pump pool heaters	All	—	4.0 COP	AHRI 1160
Unfired storage tanks	All	—	Minimum insulation requirement R-12.5 (h · ft² · °F)/Btu	(none)

For SI: °C = [(°F) - 32]/1.8, 1 British thermal unit per hour = 0.2931 W, 1 gallon = 3.785 L, 1 British thermal unit per hour per gallon = 0.078 W/L.

a. Energy factor (EF) and thermal efficiency (E_t) are minimum requirements. In the EF equation, V is the rated volume in gallons.

b. Standby loss (SL) is the maximum Btu/h based on a nominal 70°F temperature difference between stored water and ambient requirements. In the SL equation, Q is the nameplate input rate in Btu/h. In the equations for electric water heaters, V is the rated volume in gallons and V_m is the measured volume in gallons. In the SL equation for oil and gas water heaters and boilers, V is the rated volume in gallons.

c. Instantaneous water heaters with input rates below 200,000 Btu/h shall comply with these requirements where the water heater is designed to heat water to temperatures 180°F or higher.

d. Electric water heaters with an input rating of 12 kW (40,950 Btu/hr) or less that are designed to heat water to temperatures of 180°F or greater shall comply with the requirements for electric water heaters that have an input rating greater than 12 kW (40,950 Btu/h).

C404.5 Efficient heated water supply piping. Heated water supply piping shall be in accordance with Section C404.5.1 or C404.5.2. The flow rate through $^1/_4$-inch (6.4 mm) piping shall be not greater than 0.5 gpm (1.9 L/m). The flow rate through $^5/_{16}$-inch (7.9 mm) piping shall be not greater than 1 gpm (3.8 L/m). The flow rate through $^3/_8$-inch (9.5 mm) piping shall be not greater than 1.5 gpm (5.7 L/m).

C404.5.1 Maximum allowable pipe length method. The maximum allowable piping length from the nearest source of heated water to the termination of the fixture supply pipe shall be in accordance with the following. Where the piping contains more than one size of pipe, the largest size of pipe within the piping shall be used for determining the maximum allowable length of the piping in Table C404.5.1.

1. For a public lavatory faucet, use the "Public lavatory faucets" column in Table C404.5.1.
2. For all other plumbing fixtures and plumbing appliances, use the "Other fixtures and appliances" column in Table C404.5.1.

C404.5.2 Maximum allowable pipe volume method. The water volume in the piping shall be calculated in accordance with Section C404.5.2.1. Water heaters, circulating water systems and heat trace temperature maintenance systems shall be considered sources of heated water.

The volume from the nearest source of heated water to the termination of the fixture supply pipe shall be as follows:

1. For a public lavatory faucet: not more than 2 ounces (0.06 L).
2. For other plumbing fixtures or plumbing appliances; not more than 0.5 gallon (1.89 L).

C404.5.2.1 Water volume determination. The volume shall be the sum of the internal volumes of pipe, fittings, valves, meters and manifolds between the nearest source of heated water and the termination of the fixture supply pipe. The volume in the piping shall be determined from the "Volume" column in Table C404.5.1. The volume contained within fixture shutoff valves, within flexible water supply connectors to a fixture fitting and within a fixture fitting shall not be included in the water volume determination. Where heated water is supplied by a recirculating system or heat-traced piping, the volume shall include the portion of the fitting on the branch pipe that supplies water to the fixture.

C404.6 Heated-water circulating and temperature maintenance systems. Heated-water circulation systems shall be in accordance with Section C404.6.1. Heat trace temperature maintenance systems shall be in accordance with Section C404.6.2. Controls for hot water storage shall be in accordance with Section C404.6.3. Automatic controls, temperature sensors and pumps shall be *accessible*. Manual controls shall be *readily accessible*.

C404.6.1 Circulation systems. Heated-water circulation systems shall be provided with a circulation pump. The system return pipe shall be a dedicated return pipe or a cold water supply pipe. Gravity and thermo-syphon circulation systems shall be prohibited. Controls for circulating hot water system pumps shall start the pump based on the identification of a demand for hot water within the occupancy. The controls shall automatically turn off the pump when the water in the circulation loop is at the desired temperature and when there is no demand for hot water.

C404.6.2 Heat trace systems. Electric heat trace systems shall comply with IEEE 515.1. Controls for such systems shall be able to automatically adjust the energy input to the heat tracing to maintain the desired water temperature in the piping in accordance with the times when heated water is used in the occupancy. Heat trace shall be arranged to be turned off automatically when there is no hot water demand.

TABLE C404.5.1
PIPING VOLUME AND MAXIMUM PIPING LENGTHS

NOMINAL PIPE SIZE (inches)	VOLUME (liquid ounces per foot length)	MAXIMUM PIPING LENGTH (feet)	
		Public lavatory faucets	Other fixtures and appliances
$^1/_4$	0.33	6	50
$^5/_{16}$	0.5	4	50
$^3/_8$	0.75	3	50
$^1/_2$	1.5	2	43
$^5/_8$	2	1	32
$^3/_4$	3	0.5	21
$^7/_8$	4	0.5	16
1	5	0.5	13
$1^1/_4$	8	0.5	8
$1^1/_2$	11	0.5	6
2 or larger	18	0.5	4

For SI: 1 inch = 25.4 mm, 1 foot = 304.8 mm, 1 liquid ounce = 0.030 L, 1 gallon = 128 ounces.

C404.6.3 Controls for hot water storage. The controls on pumps that circulate water between a water heater and a heated-water storage tank shall limit operation of the pump from heating cycle startup to not greater than 5 minutes after the end of the cycle.

C404.7 Demand recirculation controls. A water distribution system having one or more recirculation pumps that pump water from a heated-water supply pipe back to the heated-water source through a cold-water supply pipe shall be a *demand recirculation water system*. Pumps shall have controls that comply with both of the following:

1. The control shall start the pump upon receiving a signal from the action of a user of a fixture or appliance, sensing the presence of a user of a fixture or sensing the flow of hot or tempered water to a fixture fitting or appliance.

2. The control shall limit the temperature of the water entering the cold-water piping to 104°F (40°C).

C404.8 Drain water heat recovery units. Drain water heat recovery units shall comply with CSA B55.2. Potable water-side pressure loss shall be less than 10 psi (69 kPa) at maximum design flow. For Group R occupancies, the efficiency of drain water heat recovery unit efficiency shall be in accordance with CSA B55.1.

C404.9 Energy consumption of pools and permanent spas. (Mandatory). The energy consumption of pools and permanent spas shall be controlled by the requirements in Sections C404.9.1 through C404.9.3.

C404.9.1 Heaters. The electric power to all heaters shall be controlled by a readily accessible on-off switch that is an integral part of the heater, mounted on the exterior of the heater, or external to and within 3 feet (914 mm) of the heater. Operation of such switch shall not change the setting of the heater thermostat. Such switches shall be in addition to a circuit breaker for the power to the heater. Gas-fired heaters shall not be equipped with continuously burning ignition pilots.

C404.9.2 Time switches. Time switches or other control methods that can automatically turn off and on heaters and pump motors according to a preset schedule shall be installed for heaters and pump motors. Heaters and pump motors that have built-in time switches shall be in compliance with this section.

Exceptions:

1. Where public health standards require 24-hour pump operation.

2. Pumps that operate solar- and waste-heat-recovery pool heating systems.

C404.9.3 Covers. Outdoor heated pools and outdoor permanent spas shall be provided with a vapor-retardant cover or other *approved* vapor-retardant means.

Exception: Where more than 70 percent of the energy for heating, computed over an operating season, is from site-recovered energy such as from a heat pump or solar energy source, covers or other vapor-retardant means shall not be required.

C404.10 Energy consumption of portable spas (Mandatory). The energy consumption of electric-powered portable spas shall be controlled by the requirements of APSP 14.

C404.11 Service water-heating system commissioning and completion requirements. *Service water-heating systems*, swimming pool water-heating systems, spa water-heating systems and the controls for those systems shall be commissioned and completed in accordance with Section C408.2.

SECTION C405
ELECTRICAL POWER AND LIGHTING SYSTEMS

C405.1 General (Mandatory). This section covers lighting system controls, the maximum lighting power for interior and exterior applications and electrical energy consumption.

Exception: Dwelling units within commercial buildings shall not be required to comply with Sections C405.2 through C405.5, provided that they comply with Section R404.1.

Walk-in coolers, walk-in freezers, refrigerated warehouse coolers and refrigerated warehouse freezers shall comply with Section C403.2.15 or C403.2.16.

C405.2 Lighting controls (Mandatory). Lighting systems shall be provided with controls as specified in Sections C405.2.1, C405.2.2, C405.2.3, C405.2.4 and C405.2.5.

Exceptions: Lighting controls are not required for the following:

1. Areas designated as security or emergency areas that are required to be continuously lighted.

2. Interior exit stairways, interior exit ramps and exit passageways.

3. Emergency egress lighting that is normally off.

C405.2.1 Occupant sensor controls. Occupant *sensor controls* shall be installed to control lights in the following space types:

1. Classrooms/lecture/training rooms.
2. Conference/meeting/multipurpose rooms.
3. Copy/print rooms.
4. Lounges.
5. Employee lunch and break rooms.
6. Private offices.
7. Restrooms.
8. Storage rooms.
9. Janitorial closets.
10. Locker rooms.
11. Other spaces 300 square feet (28 m^2) or less that are enclosed by floor-to-ceiling height partitions.
12. Warehouses.

C405.2.1.1 Occupant sensor control function. Occupant sensor controls in spaces other than warehouses specified in Section C405.2.1 shall comply with the following:

1. Automatically turn off lights within 30 minutes of all occupants leaving the space.

2. Be manual on or controlled to automatically turn the lighting on to not more than 50 percent power.

    Exception: Full automatic-on controls shall be permitted to control lighting in public corridors, stairways, restrooms, primary building entrance areas and lobbies, and areas where manual-on operation would endanger the safety or security of the room or building occupants.

3. Shall incorporate a *manual control* to allow occupants to turn lights off.

C405.2.1.2 Occupant sensor control function in warehouses. In warehouses, the lighting in aisleways and open areas shall be controlled with occupant sensors that automatically reduce lighting power by not less than 50 percent when the areas are unoccupied. The occupant sensors shall control lighting in each aisleway independently and shall not control lighting beyond the aisleway being controlled by the sensor.

C405.2.2 Time-switch controls. Each area of the building that is not provided with *occupant sensor controls* complying with Section C405.2.1.1 shall be provided with *time switch controls* complying with Section C405.2.2.1.

Exception: Where a *manual control* provides light reduction in accordance with Section C405.2.2.2, automatic controls shall not be required for the following:

1. *Sleeping units.*
2. Spaces where patient care is directly provided.
3. Spaces where an automatic shutoff would endanger occupant safety or security.
4. Lighting intended for continuous operation.
5. Shop and laboratory classrooms.

C405.2.2.1 Time-switch control function. Each space provided with *time-switch controls* shall also be provided with a *manual control* for light reduction in accordance with Section C405.2.2.2. Time-switch *controls* shall include an override switching device that complies with the following:

1. Have a minimum 7-day clock.
2. Be capable of being set for seven different day types per week.
3. Incorporate an automatic holiday "shutoff" feature, which turns off all controlled lighting loads for at least 24 hours and then resumes normally scheduled operations.
4. Have program backup capabilities, which prevent the loss of program and time settings for at least 10 hours, if power is interrupted.

5. Include an override switch that complies with the following:

    5.1. The override switch shall be a manual control.

    5.2. The override switch, when initiated, shall permit the controlled lighting to remain on for not more than 2 hours.

    5.3. Any individual override switch shall control the lighting for an area not larger than 5,000 square feet (465 m^2).

Exceptions:

1. Within malls, arcades, auditoriums, single-tenant retail spaces, industrial facilities and arenas:

    1.1. The time limit shall be permitted to be greater than 2 hours, provided that the override switch is a captive key device.

    1.2. The area controlled by the override switch is permitted to be greater than 5,000 square feet (465 m^2), but shall not be greater than 20,000 square feet (1860 m^2).

2. Where provided with *manual control*, the following areas are not required to have light reduction control:

    2.1. Spaces that have only one luminaire with a rated power of less than 100 watts.

    2.2. Spaces that use less than 0.6 watts per square foot (6.5 W/m^2).

    2.3. Corridors, equipment rooms, public lobbies, electrical or mechanical rooms.

C405.2.2.2 Light-reduction controls. Spaces required to have light-reduction controls shall have a *manual control* that allows the occupant to reduce the connected lighting load in a reasonably uniform illumination pattern by at least 50 percent. Lighting reduction shall be achieved by one of the following or another *approved* method:

1. Controlling all lamps or luminaires.
2. Dual switching of alternate rows of luminaires, alternate luminaires or alternate lamps.
3. Switching the middle lamp luminaires independently of the outer lamps.
4. Switching each luminaire or each lamp.

Exception: Light reduction controls are not required in *daylight zones* with *daylight responsive controls* complying with Section C405.2.3.

C405.2.2.3 Manual controls. *Manual controls* for lights shall comply with the following:

1. Shall be readily accessible to occupants.

2. Shall be located where the controlled lights are visible, or shall identify the area served by the lights and indicate their status.

C405.2.3 Daylight-responsive controls. *Daylight-responsive controls* complying with Section C405.2.3.1 shall be provided to control the electric lights within *daylight zones* in the following spaces:

1. Spaces with a total of more than 150 watts of *general lighting* within sidelight *daylight zones* complying with Section C405.2.3.2. *General lighting* does not include lighting that is required to have specific application control in accordance with Section C405.2.4.

2. Spaces with a total of more than 150 watts of *general lighting* within toplight *daylight zones* complying with Section C405.2.3.3.

Exceptions: Daylight responsive controls are not required for the following:

1. Spaces in health care facilities where patient care is directly provided.

2. Dwelling units and sleeping units.

3. Lighting that is required to have specific application control in accordance with Section C405.2.4.

4. Sidelight daylight zones on the first floor above grade in Group A-2 and Group M occupancies.

C405.2.3.1 Daylight-responsive control function. Where required, *daylight-responsive controls* shall be provided within each space for control of lights in that space and shall comply with all of the following:

1. Lights in toplight *daylight zones* in accordance with Section C405.2.3.3 shall be controlled independently of lights in sidelight *daylight zones* in accordance with Section C405.2.3.2.

2. *Daylight responsive controls* within each space shall be configured so that they can be calibrated from within that space by authorized personnel.

3. Calibration mechanisms shall be r*eadily accessible.*

4. Where located in offices, classrooms, laboratories and library reading rooms, *daylight responsive controls* shall dim lights continuously from full light output to 15 percent of full light output or lower.

5. *Daylight responsive controls* shall be capable of a complete shutoff of all controlled lights.

6. Lights in sidelight *daylight zones* in accordance with Section C405.2.3.2 facing different cardinal orientations [i.e., within 45 degrees (0.79 rad) of due north, east, south, west] shall be controlled independently of each other.

Exception: Up to 150 watts of lighting in each space is permitted to be controlled together with lighting in a daylight zone facing a different cardinal orientation.

C405.2.3.2 Sidelight daylight zone. The sidelight *daylight zone* is the floor area adjacent to vertical *fenestration* which complies with all of the following:

1. Where the fenestration is located in a wall, the daylight zone shall extend laterally to the nearest full-height wall, or up to 1.0 times the height from the floor to the top of the fenestration, and longitudinally from the edge of the fenestration to the nearest full-height wall, or up to 2 feet (610 mm), whichever is less, as indicated in Figure C405.2.3.2(1).

2. Where the *fenestration* is located in a rooftop monitor, the *daylight zone* shall extend laterally to the nearest obstruction that is taller than 0.7 times the ceiling height, or up to 1.0 times the height from the floor to the bottom of the *fenes-*

(a) Section view
(b) Plan view of daylight zone under a rooftop monitor

**FIGURE C405.2.3.2(1)
DAYLIGHT ZONE ADJACENT TO FENESTRATION IN A WALL**

tration, whichever is less, and longitudinally from the edge of the *fenestration* to the nearest obstruction that is taller than 0.7 times the ceiling height, or up to 0.25 times the height from the floor to the bottom of the *fenestration*, whichever is less, as indicated in Figures C405.2.3.2(2) and C405.2.3.2(3).

3. The area of the *fenestration* is not less than 24 square feet (2.23 m^2).

4. The distance from the *fenestration* to any building or geological formation which would block access to daylight is greater than the height from the bottom of the *fenestration* to the top of the building or geologic formation.

5. Where located in existing buildings, the *visible transmittance* of the *fenestration* is not less than 0.20.

C405.2.3.3 Toplight daylight zone. The toplight *daylight zone* is the floor area underneath a roof fenestration assembly which complies with all of the following:

1. The *daylight zone* shall extend laterally and longitudinally beyond the edge of the roof *fenestration* assembly to the nearest obstruction that is taller than 0.7 times the ceiling height, or up to 0.7 times the ceiling height, whichever is less, as indicated in Figure C405.2.3.3.

2. No building or geological formation blocks direct sunlight from hitting the roof *fenestration* assembly at the peak solar angle on the summer solstice.

3. Where located in existing buildings, the product of the *visible transmittance* of the roof *fenestration* assembly and the area of the rough opening of the roof *fenestration* assembly divided by the area of the *daylight zone* is not less than 0.008.

FIGURE C405.2.3.2(2)
DAYLIGHT ZONE UNDER A ROOFTOP MONITOR

FIGURE C405.2.3.2(3)
DAYLIGHT ZONE UNDER A SLOPED ROOFTOP MONITOR

COMMERCIAL ENERGY EFFICIENCY

(a) Section view
(b) Plan view of daylight zone under a roof fenestration assembly

**FIGURE C405.2.3.3
DAYLIGHT ZONE UNDER A ROOF FENESTRATION ASSEMBLY**

C405.2.4 Specific application controls. Specific application controls shall be provided for the following:

1. Display and accent light shall be controlled by a dedicated control that is independent of the controls for other lighting within the room or space.

2. Lighting in cases used for display case purposes shall be controlled by a dedicated control that is independent of the controls for other lighting within the room or space.

3. Hotel and motel sleeping units and guest suites shall have a master control device that is capable of automatically switching off all installed luminaires and switched receptacles within 20 minutes after all occupants leave the room.

   Exception: Lighting and switched receptacles controlled by captive key systems.

4. Supplemental task lighting, including permanently installed under-shelf or under-cabinet lighting, shall have a control device integral to the luminaires or be controlled by a wall-mounted control device provided that the control device is readily accessible.

5. Lighting for nonvisual applications, such as plant growth and food warming, shall be controlled by a dedicated control that is independent of the controls for other lighting within the room or space.

6. Lighting equipment that is for sale or for demonstrations in lighting education shall be controlled by a dedicated control that is independent of the controls for other lighting within the room or space.

C405.2.5 Exterior lighting controls. Lighting for exterior applications other than emergency lighting that is intended to be automatically off during building operation, lighting specifically required to meet health and life safety requirements or decorative gas lighting systems shall:

1. Be provided with a control that automatically turns off the lighting as a function of available daylight.

2. Where lighting the building façade or landscape, the lighting shall have controls that automatically shut off the lighting as a function of dawn/dusk and a set opening and closing time.

3. Where not covered in Item 2, the lighting shall have controls configured to automatically reduce the connected lighting power by not less than 30 percent from not later than midnight to 6 a.m., from one hour after business closing to one hour before business opening or during any period when activity has not been detected for a time of longer than 15 minutes.

All time switches shall be able to retain programming and the time setting during loss of power for a period of at least 10 hours.

Exception: Lighting for covered vehicle entrances or exits from buildings or parking structures where required for safety, security or eye adaptation.

C405.3 Exit signs (Mandatory). Internally illuminated exit signs shall not be more than 5 watts per side.

C405.4 Interior lighting power requirements (Prescriptive). A building complies with this section where its total connected lighting power calculated under Section C405.4.1 is not greater than the interior lighting power calculated under Section C405.4.2.

C405.4.1 Total connected interior lighting power. The total connected interior lighting power shall be determined in accordance with Equation 4-9.

$TCLP = [SL + LV + LTPB + \text{Other}]$ **(Equation 4-9)**

where:

$TCLP$ = Total connected lighting power (watts).

SL = Labeled wattage of luminaires for screw-in lamps.

LV = Wattage of the transformer supplying low-voltage lighting.

$LTPB$ = Wattage of line-voltage lighting tracks and plug-in busways as the specified wattage of the luminaires, but at least 30 W/lin. ft. (100 W/lin m), or the wattage limit of the system's circuit breaker, or the wattage limit of other permanent current-limiting devices on the system.

Other = The wattage of all other luminaires and lighting sources not covered previously and associated with interior lighting verified by data supplied by the manufacturer or other *approved* sources.

Exceptions:

1. The connected power associated with the following lighting equipment is not included in calculating total connected lighting power.

    1.1. Professional sports arena playing field lighting.

    1.2. Lighting in sleeping units, provided that the lighting complies with Section R404.1.

    1.3. Emergency lighting automatically off during normal building operation.

    1.4. Lighting in spaces specifically designed for use by occupants with special lighting needs, including those with visual impairment and other medical and age-related issues.

    1.5. Lighting in interior spaces that have been specifically designated as a registered interior historic landmark.

    1.6. Casino gaming areas.

    1.7. Mirror lighting in dressing rooms.

2. Lighting equipment used for the following shall be exempt provided that it is in addition to general lighting and is controlled by an independent control device:

    2.1. Task lighting for medical and dental purposes.

    2.2. Display lighting for exhibits in galleries, museums and monuments.

3. Lighting for theatrical purposes, including performance, stage, film production and video production.

4. Lighting for photographic processes.

5. Lighting integral to equipment or instrumentation and installed by the manufacturer.

6. Task lighting for plant growth or maintenance.

7. Advertising signage or directional signage.

8. In restaurant buildings and areas, lighting for food warming or integral to food preparation equipment.

9. Lighting equipment that is for sale.

10. Lighting demonstration equipment in lighting education facilities.

11. Lighting *approved* because of safety or emergency considerations, inclusive of exit lights.

12. Lighting integral to both open and glass-enclosed refrigerator and freezer cases.

13. Lighting in retail display windows, provided the display area is enclosed by ceiling-height partitions.

14. Furniture-mounted supplemental task lighting that is controlled by automatic shutoff.

15. Exit signs.

C405.4.2 Interior lighting power. The total interior lighting power allowance (watts) is determined according to Table C405.4.2(1) using the Building Area Method, or Table C405.4.2(2) using the Space-by-Space Method, for all areas of the building covered in this permit.

C405.4.2.1 Building Area Method. For the Building Area Method, the interior lighting power allowance is the floor area for each building area type listed in Table C405.4.2(1) times the value from Table C405.4.2(1) for that area. For the purposes of this method, an "area" shall be defined as all contiguous spaces that accommodate or are associated with a single building area type, as listed in Table C405.4.2(1). Where this method is used to calculate the total interior lighting power for an entire building, each building area type shall be treated as a separate area.

C405.4.2.2 Space-by-Space Method. For the Space-by-Space Method, the interior lighting power allowance is determined by multiplying the floor area of each space times the value for the space type in Table C405.4.2(2) that most closely represents the proposed use of the space, and then summing the lighting power allowances for all spaces. Trade-offs among spaces are permitted.

C405.4.2.2.1 Additional interior lighting power. Where using the Space-by-Space Method, an increase in the interior lighting power allowance is permitted for specific lighting functions. Additional power shall be permitted only where the specified lighting is installed and automatically controlled separately from the general lighting, to be turned off during nonbusiness hours. This additional power shall be used only for the specified luminaires and shall not be used for any other purpose. An increase in the interior lighting power allowance is permitted in the following cases:

1. For lighting equipment to be installed in sales areas specifically to highlight merchandise, the additional lighting power shall be determined in accordance with Equation 4-10.

    Additional interior lighting power allowance = 500 watts + (Retail Area 1 · 0.6 W/ft^2) + (Retail Area 2 · 0.6 W/ft^2) + (Retail Area 3 · 1.4 W/ft^2) + (Retail Area 4 · 2.5 W/ft^2)

    (Equation 4-10)

where:

Retail Area 1 = The floor area for all products not listed in Retail Area 2, 3 or 4.

Retail Area 2 = The floor area used for the sale of vehicles, sporting goods and small electronics.

Retail Area 3 = The floor area used for the sale of furniture, clothing, cosmetics and artwork.

Retail Area 4 = The floor area used for the sale of jewelry, crystal and china.

Exception: Other merchandise categories are permitted to be included in Retail Areas 2 through 4, provided that justification documenting the need for additional lighting power based on visual inspection, contrast, or other critical display is *approved* by the code official.

2. For spaces in which lighting is specified to be installed in addition to the general lighting for the purpose of decorative appearance or for highlighting art or exhibits, provided that the additional lighting power shall be not more than 1.0 w/ft^2 (10.7 w/m^2) of such spaces.

TABLE C405.4.2(1)
INTERIOR LIGHTING POWER ALLOWANCES:
BUILDING AREA METHOD

BUILDING AREA TYPE	LPD (w/ft^2)
Automotive facility	0.80
Convention center	1.01
Courthouse	1.01
Dining: bar lounge/leisure	1.01
Dining: cafeteria/fast food	0.9
Dining: family	0.95
Dormitory	0.57
Exercise center	0.84
Fire station	0.67
Gymnasium	0.94
Health care clinic	0.90
Hospital	1.05
Hotel/Motel	0.87
Library	1.19
Manufacturing facility	1.17
Motion picture theater	0.76
Multifamily	0.51
Museum	1.02
Office	0.82
Parking garage	0.21
Penitentiary	0.81
Performing arts theater	1.39
Police station	0.87
Post office	0.87
Religious building	1.0
Retail	1.26
School/university	0.87
Sports arena	0.91
Town hall	0.89
Transportation	0.70
Warehouse	0.66
Workshop	1.19

TABLE C405.4.2(2)
INTERIOR LIGHTING POWER ALLOWANCES:
SPACE-BY-SPACE METHOD

COMMON SPACE TYPES[a]	LPD (watts/sq.ft)
Atrium	
Less than 40 feet in height	0.03 per foot in total height
Greater than 40 feet in height	0.40 + 0.02 per foot in total height
Audience seating area	
In an auditorium	0.63
In a convention center	0.82
In a gymnasium	0.65
In a motion picture theater	1.14
In a penitentiary	0.28
In a performing arts theater	2.43
In a religious building	1.53
In a sports arena	0.43
Otherwise	0.43
Banking activity area	1.01
Breakroom (See Lounge/Breakroom)	
Classroom/lecture hall/training room	
In a penitentiary	1.34
Otherwise	1.24
Conference/meeting/multipurpose room	1.23
Copy/print room	0.72
Corridor	
In a facility for the visually impaired (and not used primarily by the staff)[b]	0.92
In a hospital	0.79
In a manufacturing facility	0.41
Otherwise	0.66
Courtroom	1.72
Computer room	1.71
Dining area	
In a penitentiary	0.96
In a facility for the visually impaired (and not used primarily by the staff)[b]	1.9
In bar/lounge or leisure dining	1.07
In cafeteria or fast food dining	0.65
In family dining	0.89
Otherwise	0.65
Electrical/mechanical room	0.95
Emergency vehicle garage	0.56

(continued)

TABLE C405.4.2(2)—continued
INTERIOR LIGHTING POWER ALLOWANCES: SPACE-BY-SPACE METHOD

COMMON SPACE TYPES[a]	LPD (watts/sq.ft)
Food preparation area	1.21
Guest room	0.47
Laboratory	
In or as a classroom	1.43
Otherwise	1.81
Laundry/washing area	0.6
Loading dock, interior	0.47
Lobby	
In a facility for the visually impaired (and not used primarily by the staff)[b]	1.8
For an elevator	0.64
In a hotel	1.06
In a motion picture theater	0.59
In a performing arts theater	2.0
Otherwise	0.9
Locker room	0.75
Lounge/breakroom	
In a healthcare facility	0.92
Otherwise	0.73
Office	
Enclosed	1.11
Open plan	0.98
Parking area, interior	0.19
Pharmacy area	1.68
Restroom	
In a facility for the visually impaired (and not used primarily by the staff[b]	1.21
Otherwise	0.98
Sales area	1.59
Seating area, general	0.54
Stairway (See space containing stairway)	
Stairwell	0.69
Storage room	0.63
Vehicular maintenance area	0.67
Workshop	1.59
BUILDING TYPE SPECIFIC SPACE TYPES[a]	**LPD (watts/sq.ft)**
Facility for the visually impaired[b]	
In a chapel (and not used primarily by the staff)	2.21
In a recreation room (and not used primarily by the staff)	2.41
Automotive (See Vehicular Maintenance Area above)	
Convention Center—exhibit space	1.45
Dormitory—living quarters	0.38
Fire Station—sleeping quarters	0.22
Gymnasium/fitness center	
In an exercise area	0.72
In a playing area	1.2

(continued)

TABLE C405.4.2(2)—continued
INTERIOR LIGHTING POWER ALLOWANCES: SPACE-BY-SPACE METHOD

BUILDING TYPE SPECIFIC SPACE TYPES[a]	LPD (watts/sq.ft)
healthcare facility	
In an exam/treatment room	1.66
In an imaging room	1.51
In a medical supply room	0.74
In a nursery	0.88
In a nurse's station	0.71
In an operating room	2.48
In a patient room	0.62
In a physical therapy room	0.91
In a recovery room	1.15
Library	
In a reading area	1.06
In the stacks	1.71
Manufacturing facility	
In a detailed manufacturing area	1.29
In an equipment room	0.74
In an extra high bay area (greater than 50' floor-to-ceiling height)	1.05
In a high bay area (25-50' floor-to-ceiling height)	1.23
In a low bay area (less than 25' floor-to-ceiling height)	1.19
Museum	
In a general exhibition area	1.05
In a restoration room	1.02
Performing arts theater—dressing room	0.61
Post Office—Sorting Area	0.94
Religious buildings	
In a fellowship hall	0.64
In a worship/pulpit/choir area	1.53
Retail facilities	
In a dressing/fitting room	0.71
In a mall concourse	1.1
Sports arena—playing area	
For a Class I facility	3.68
For a Class II facility	2.4
For a Class III facility	1.8
For a Class IV facility	1.2
Transportation facility	
In a baggage/carousel area	0.53
In an airport concourse	0.36
At a terminal ticket counter	0.8
Warehouse—storage area	
For medium to bulky, palletized items	0.58
For smaller, hand-carried items	0.95

a. In cases where both a common space type and a building area specific space type are listed, the building area specific space type shall apply
b. A 'Facility for the Visually Impaired' is a facility that is licensed or will be licensed by local or state authorities for senior long-term care, adult daycare, senior support or people with special visual needs.

C405.5 Exterior lighting (Mandatory). Where the power for exterior lighting is supplied through the energy service to the building, all exterior lighting shall comply with Section C405.5.1.

> **Exception:** Where *approved* because of historical, safety, signage or emergency considerations.

C405.5.1 Exterior building lighting power. The total exterior lighting power allowance for all exterior building applications is the sum of the base site allowance plus the individual allowances for areas that are to be illuminated and are permitted in Table C405.5.1(2) for the applicable lighting zone. Trade-offs are allowed only among exterior lighting applications listed in Table C405.5.1(2), in the Tradable Surfaces section. The lighting zone for the building exterior is determined from Table C405.5.1(1) unless otherwise specified by the local jurisdiction.

> **Exception:** Lighting used for the following exterior applications is exempt where equipped with a control device independent of the control of the nonexempt lighting:
> 1. Specialized signal, directional and marker lighting associated with transportation.
> 2. Advertising signage or directional signage.
> 3. Integral to equipment or instrumentation and is installed by its manufacturer.
> 4. Theatrical purposes, including performance, stage, film production and video production.
> 5. Athletic playing areas.
> 6. Temporary lighting.
> 7. Industrial production, material handling, transportation sites and associated storage areas.
> 8. Theme elements in theme/amusement parks.
> 9. Used to highlight features of public monuments and registered historic landmark structures or buildings.

TABLE C405.5.1(1)
EXTERIOR LIGHTING ZONES

LIGHTING ZONE	DESCRIPTION
1	Developed areas of national parks, state parks, forest land, and rural areas
2	Areas predominantly consisting of residential zoning, neighborhood business districts, light industrial with limited nighttime use and residential mixed-use areas
3	All other areas not classified as lighting zone 1, 2 or 4
4	High-activity commercial districts in major metropolitan areas as designated by the local land use planning authority

C405.6 Electrical energy consumption (Mandatory). Each dwelling unit located in a Group R-2 building shall have a separate electrical meter.

C405.7 Electrical transformers (Mandatory). Electric transformers shall meet the minimum efficiency requirements of Table C405.7 as tested and rated in accordance with the test procedure listed in DOE 10 CFR 431. The efficiency shall be verified through certification under an approved certification program or, where a certification program does not exist, the equipment efficiency ratings shall be supported by data furnished by the transformer manufacturer.

> **Exceptions:** The following transformers are exempt:
> 1. Transformers that meet the *Energy Policy Act of 2005* exclusions based on the DOE 10 CFR 431 definition of special purpose applications.
> 2. Transformers that meet the *Energy Policy Act of 2005* exclusions that are not to be used in general purpose applications based on information provided in DOE 10 CFR 431.
> 3. Transformers that meet the *Energy Policy Act of 2005* exclusions with multiple voltage taps where the highest tap is at least 20 percent more than the lowest tap.
> 4. Drive transformers.
> 5. Rectifier transformers.
> 6. Auto-transformers.
> 7. Uninterruptible power system transformers.
> 8. Impedance transformers.
> 9. Regulating transformers.
> 10. Sealed and nonventilating transformers.
> 11. Machine tool transformers.
> 12. Welding transformers.
> 13. Grounding transformers.
> 14. Testing transformers.

C405.8 Electrical motors (Mandatory). Electric motors shall meet the minimum efficiency requirements of Tables C405.8(1) through C405.8(4) when tested and rated in accordance with the DOE 10 CFR 431. The efficiency shall be verified through certification under an approved certification program or, where a certification program does not exist, the equipment efficiency ratings shall be supported by data furnished by the motor manufacturer.

C405.9 Vertical and horizontal transportation systems and equipment. Vertical and horizontal transportation systems and equipment shall comply with this section.

C405.9.1 Elevator cabs. For the luminaires in each elevator cab, not including signals and displays, the sum of the lumens divided by the sum of the watts shall be not less than 35 lumens per watt. Ventilation fans in elevators that do not have their own air-conditioning system shall not consume more than 0.33 watts/cfm at the maximum rated speed of the fan. Controls shall be provided that will de-energize ventilation fans and lighting systems when the elevator is stopped, unoccupied and with its doors closed for over 15 minutes.

C405.9.2 Escalators and moving walks. Escalators and moving walks shall comply with ASME A17.1/CSA B44 and shall have automatic controls configured to reduce speed to the minimum permitted speed in accordance with

ASME A17.1/CSA B44 or applicable local code when not conveying passengers.

C405.9.2.1 Regenerative drive. An escalator designed either for one-way down operation only or for reversible operation shall have a variable frequency regenerative drive that supplies electrical energy to the building electrical system when the escalator is loaded with passengers whose combined weight exceeds 750 pounds (340 kg).

TABLE C405.5.1(2)
INDIVIDUAL LIGHTING POWER ALLOWANCES FOR BUILDING EXTERIORS

		LIGHTING ZONES			
		Zone 1	Zone 2	Zone 3	Zone 4
Base Site Allowance (Base allowance is usable in tradable or nontradable surfaces.)		500 W	600 W	750 W	1300 W
	Uncovered Parking Areas				
	Parking areas and drives	0.04 W/ft^2	0.06 W/ft^2	0.10 W/ft^2	0.13 W/ft^2
	Building Grounds				
	Walkways less than 10 feet wide	0.7 W/linear foot	0.7 W/linear foot	0.8 W/linear foot	1.0 W/linear foot
	Walkways 10 feet wide or greater, plaza areas special feature areas	0.14 W/ft^2	0.14 W/ft^2	0.16 W/ft^2	0.2 W/ft^2
	Stairways	0.75 W/ft^2	1.0 W/ft^2	1.0 W/ft^2	1.0 W/ft^2
Tradable Surfaces (Lighting power densities for uncovered parking areas, building grounds, building entrances and exits, canopies and overhangs and outdoor sales areas are tradable.)	Pedestrian tunnels	0.15 W/ft^2	0.15 W/ft^2	0.2 W/ft^2	0.3 W/ft^2
	Building Entrances and Exits				
	Main entries	20 W/linear foot of door width	20 W/linear foot of door width	30 W/linear foot of door width	30 W/linear foot of door width
	Other doors	20 W/linear foot of door width	20 W/linear foot of door width	20 W/linear foot of door width	20 W/linear foot of door width
	Entry canopies	0.25 W/ft^2	0.25 W/ft^2	0.4 W/ft^2	0.4 W/ft^2
	Sales Canopies				
	Free-standing and attached	0.6 W/ft^2	0.6 W/ft^2	0.8 W/ft^2	1.0 W/ft^2
	Outdoor Sales				
	Open areas (including vehicle sales lots)	0.25 W/ft^2	0.25 W/ft^2	0.5 W/ft^2	0.7 W/ft^2
	Street frontage for vehicle sales lots in addition to "open area" allowance	No allowance	10 W/linear foot	10 W/linear foot	30 W/linear foot
Nontradable Surfaces (Lighting power density calculations for the following applications can be used only for the specific application and cannot be traded between surfaces or with other exterior lighting. The following allowances are in addition to any allowance otherwise permitted in the "Tradable Surfaces" section of this table.)	Building facades	No allowance	0.075 W/ft^2 of gross above-grade wall area	0.113 W/ft^2 of gross above-grade wall area	0.15 W/ft^2 of gross above-grade wall area
	Automated teller machines (ATM) and night depositories	270 W per location plus 90 W per additional ATM per location	270 W per location plus 90 W per additional ATM per location	270 W per location plus 90 W per additional ATM per location	270 W per location plus 90 W per additional ATM per location
	Entrances and gatehouse inspection stations at guarded facilities	0.75 W/ft^2 of covered and uncovered area	0.75 W/ft^2 of covered and uncovered area	0.75 W/ft^2 of covered and uncovered area	0.75 W/ft^2 of covered and uncovered area
	Loading areas for law enforcement, fire, ambulance and other emergency service vehicles	0.5 W/ft^2 of covered and uncovered area	0.5 W/ft^2 of covered and uncovered area	0.5 W/ft^2 of covered and uncovered area	0.5 W/ft^2 of covered and uncovered area
	Drive-up windows/doors	400 W per drive-through	400 W per drive-through	400 W per drive-through	400 W per drive-through
	Parking near 24-hour retail entrances	800 W per main entry	800 W per main entry	800 W per main entry	800 W per main entry

For SI: 1 foot = 304.8 mm, 1 watt per square foot = W/0.0929 m^2.
W = watts.

COMMERCIAL ENERGY EFFICIENCY

TABLE C405.7
MINIMUM NOMINAL EFFICIENCY LEVELS FOR 10 CFR 431 LOW-VOLTAGE DRY-TYPE DISTRIBUTION TRANSFORMERS

SINGLE-PHASE TRANSFORMERS		THREE-PHASE TRANSFORMERS	
kVA[a]	Efficiency (%)[b]	kVA[a]	Efficiency (%)[b]
15	97.7	15	97.0
25	98.0	30	97.5
37.5	98.2	45	97.7
50	98.3	75	98.0
75	98.5	112.5	98.2
100	98.6	150	98.3
167	98.7	225	98.5
250	98.8	300	98.6
333	98.9	500	98.7
		750	98.8
		1000	98.9

a. kiloVolt-Amp rating.
b. Nominal efficiencies shall be established in accordance with the DOE 10 CFR 431 test procedure for low-voltage dry-type transformers.

TABLE C405.8(1)
MINIMUM NOMINAL FULL-LOAD EFFICIENCY FOR 60 HZ NEMA
GENERAL PURPOSE ELECTRIC MOTORS (SUBTYPE I) RATED 600 VOLTS OR LESS (Random Wound)[a]

MOTOR HORSEPOWER	NUMBER OF POLES	OPEN DRIP-PROOF MOTORS			TOTALLY ENCLOSED FAN-COOLED MOTORS		
		2	4	6	2	4	6
	Synchronous Speed (RPM)	3600	1800	1200	3600	1800	1200
1		77.0	85.5	82.5	77.0	85.5	82.5
1.5		84.0	86.5	86.5	84.0	86.5	87.5
2		85.5	86.5	87.5	85.5	86.5	88.5
3		85.5	89.5	88.5	86.5	89.5	89.5
5		86.5	89.5	89.5	88.5	89.5	89.5
7.5		88.5	91.0	90.2	89.5	91.7	91.0
10		89.5	91.7	91.7	90.2	91.7	91.0
15		90.2	93.0	91.7	91.0	92.4	91.7
20		91.0	93.0	92.4	91.0	93.0	91.7
25		91.7	93.6	93.0	91.7	93.6	93.0
30		91.7	94.1	93.6	91.7	93.6	93.0
40		92.4	94.1	94.1	92.4	94.1	94.1
50		93.0	94.5	94.1	93.0	94.5	94.1
60		93.6	95.0	94.5	93.6	95.0	94.5
75		93.6	95.0	94.5	93.6	95.4	94.5
100		93.6	95.4	95.0	94.1	95.4	95.0
125		94.1	95.4	95.0	95.0	95.4	95.0
150		94.1	95.8	95.4	95.0	95.8	95.8
200		95.0	95.8	95.4	95.4	96.2	95.8
250		95.0	95.8	95.4	95.8	96.2	95.8
300		95.4	95.8	95.4	95.8	96.2	95.8
350		95.4	95.8	95.4	95.8	96.2	95.8
400		95.8	95.8	95.8	95.8	96.2	95.8
450		95.8	96.2	96.2	95.8	96.2	95.8
500		95.8	96.2	96.2	95.8	96.2	95.8

a. Nominal efficiencies shall be established in accordance with DOE 10 CFR 431.

COMMERCIAL ENERGY EFFICIENCY

TABLE C405.8(2)
MINIMUM NOMINAL FULL-LOAD EFFICIENCY OF
GENERAL PURPOSE ELECTRIC MOTORS (SUBTYPE II) AND ALL DESIGN B MOTORS GREATER THAN 200 HORSEPOWER[a]

MOTOR HORSEPOWER	NUMBER OF POLES	OPEN DRIP-PROOF MOTORS				TOTALLY ENCLOSED FAN-COOLED MOTORS			
		2	4	6	8	2	4	6	8
	Synchronous Speed (RPM)	3600	1800	1200	900	3600	1800	1200	900
1		NR	82.5	80.0	74.0	75.5	82.5	80.0	74.0
1.5		82.5	84.0	84.0	75.5	82.5	84.0	85.5	77.0
2		84.0	84.0	85.5	85.5	84.0	84.0	86.5	82.5
3		84.0	86.5	86.5	86.5	85.5	87.5	87.5	84.0
5		85.5	87.5	87.5	87.5	87.5	87.5	87.5	84.0
7.5		87.5	88.5	88.5	88.5	88.5	89.5	89.5	85.5
10		88.5	89.5	90.2	89.5	89.5	89.5	89.5	88.5
15		89.5	91.0	90.2	89.5	90.2	91.0	90.2	88.5
20		90.2	91.0	91.0	90.2	90.2	91.0	90.2	89.5
25		91.0	91.7	91.7	90.2	91.0	92.4	91.7	89.5
30		91.0	92.4	92.4	91.0	91.0	92.4	91.7	91.0
40		91.7	93.0	93.0	91.0	91.7	93.0	93.0	91.0
50		92.4	93.0	93.0	91.7	92.4	93.0	93.0	91.7
60		93.0	93.6	93.6	92.4	93.0	93.6	93.6	91.7
75		93.0	94.1	93.6	93.6	93.0	94.1	93.6	93.0
100		93.0	94.1	94.1	93.6	93.6	94.5	94.1	93.0
125		93.6	94.5	94.1	93.6	94.5	94.5	94.1	93.6
150		93.6	95.0	94.5	93.6	94.5	95.0	95.0	93.6
200		94.5	95.0	94.5	93.6	95.0	95.0	95.0	94.1
250		94.5	95.4	95.4	94.5	95.4	95.0	95.0	94.5
300		95.0	95.4	95.4	NR	95.4	95.4	95.0	NR
350		95.0	95.4	95.4	NR	95.4	95.4	95.0	NR
400		95.4	95.4	NR	NR	95.4	95.4	NR	NR
450		95.8	95.8	NR	NR	95.4	95.4	NR	NR
500		95.8	95.8	NR	NR	95.4	95.8	NR	NR

NR = No requirement.

a. Nominal efficiencies shall be established in accordance with DOE 10 CFR 431.

TABLE C405.8(3)
MINIMUM AVERAGE FULL LOAD EFICIENCY POLYPHASE SMALL ELECTRIC MOTORS[a]

MOTOR HORSEPOWER	OPEN MOTORS			
	Number of Poles	2	4	6
	Synchronous Speed (RPM)	3600	1800	1200
0.25		65.6	69.5	67.5
0.33		69.5	73.4	71.4
0.50		73.4	78.2	75.3
0.75		76.8	81.1	81.7
1		77.0	83.5	82.5
1.5		84.0	86.5	83.8
2		85.5	86.5	N/A
3		85.5	86.9	N/A

a. Average full load efficiencies shall be established in accordance with 10 CFR 431.

TABLE C405.8(4)
MINIMUM AVERAGE FULL LOAD EFFICIENCY FOR
CAPACITOR-START CAPACITOR-RUN AND CAPACITOR-START INDUCTION-RUN SMALL ELECTRIC MOTORS[a]

MOTOR HORSEPOWER	OPEN MOTORS			
	Number of Poles	2	4	6
	Synchronous Speed (RPM)	3600	1800	1200
0.25		66.6	68.5	62.2
0.33		70.5	72.4	66.6
0.50		72.4	76.2	76.2
0.75		76.2	81.8	80.2
1		80.4	82.6	81.1
1.5		81.5	83.8	N/A
2		82.9	84.5	N/A
3		84.1	N/A	N/A

a. Average full load efficiencies shall be established in accordance with 10 CFR 431.

SECTION C406
ADDITIONAL EFFICIENCY PACKAGE OPTIONS

C406.1 Requirements. Buildings shall comply with at least one of the following:

1. More efficient HVAC performance in accordance with Section C406.2.
2. Reduced lighting power density system in accordance with Section C406.3.
3. Enhanced lighting controls in accordance with Section C406.4.
4. On-site supply of renewable energy in accordance with Section C406.5.
5. Provision of a dedicated outdoor air system for certain HVAC equipment in accordance with Section C406.6.
6. High-efficiency service water heating in accordance with Section C406.7.

C406.1.1 Tenant spaces. Tenant spaces shall comply with Section C406.2, C406.3, C406.4, C406.6 or C406.7. Alternatively, tenant spaces shall comply with Section C406.5 where the entire building is in compliance.

C406.2 More efficient HVAC equipment performance. Equipment shall exceed the minimum efficiency requirements listed in Tables C403.2.3(1) through C403.2.3(7) by 10 percent, in addition to the requirements of Section C403. Where multiple performance requirements are provided, the equipment shall exceed all requirements by 10 percent. *Variable refrigerant flow systems* shall exceed the energy efficiency provisions of ANSI/ASHRAE/IES 90.1 by 10 percent. Equipment not listed in Tables C403.2.3(1) through C403.2.3(7) shall be limited to 10 percent of the total building system capacity.

C406.3 Reduced lighting power density. The total interior lighting power (watts) of the building shall be determined by using 90 percent of the lighting power values specified in Table C405.4.2(1) times the floor area for the building types, or by using 90 percent of the interior lighting power allowance calculated by the Space-by-Space Method in Section C405.4.2.

C406.4 Enhanced digital lighting controls. Interior lighting in the building shall have the following enhanced lighting controls that shall be located, scheduled and operated in accordance with Section C405.2.2.

1. Luminaires shall be capable of continuous dimming.
2. Luminaires shall be capable of being addressed individually. Where individual addressability is not available for the luminaire class type, a controlled group of not more than four luminaires shall be allowed.
3. Not more than eight luminaires shall be controlled together in a *daylight zone*.
4. Fixtures shall be controlled through a digital control system that includes the following function:
    4.1. Control reconfiguration based on digital addressability.
    4.2. Load shedding.
    4.3. Individual user control of overhead general illumination in open offices.
    4.4. Occupancy sensors shall be capable of being reconfigured through the digital control system.
5. Construction documents shall include submittal of a Sequence of Operations, including a specification outlining each of the functions in Item 4 of this section.
6. Functional testing of lighting controls shall comply with Section C408.

C406.5 On-site renewable energy. Total minimum ratings of on-site renewable energy systems shall comply with one of the following:

1. Provide not less than 0.50 watts per square foot (5.4 W/m^2) of conditioned floor area.
2. Provide not less than 3 percent of the energy used within the building for building mechanical and service water heating equipment and lighting regulated in Chapter 4.

C406.6 Dedicated outdoor air system. Buildings covered by Section C403.4 shall be equipped with an independent ventilation system designed to provide not less than the minimum 100-percent outdoor air to each individual occupied space, as specified by the *International Mechanical Code*. The ventilation system shall be capable of total energy recovery. The HVAC system shall include supply-air temperature controls that automatically reset the supply-air temperature in response to representative building loads, or to outdoor air temperatures. The controls shall reset the supply-air temperature at least 25 percent of the difference between the design supply-air temperature and the design room-air temperature.

C406.7 Reduced energy use in service water heating. Buildings shall be of the following types to use this compliance method:

1. Group R-1: Boarding houses, hotels or motels.
2. Group I-2: Hospitals, psychiatric hospitals and nursing homes.
3. Group A-2: Restaurants and banquet halls or buildings containing food preparation areas.
4. Group F: Laundries.
5. Group R-2: Buildings with residential occupancies.
6. Group A-3: Health clubs and spas.
7. Buildings showing a service hot water load of 10 percent or more of total building energy loads, as shown with an energy analysis as described in Section C407.

C406.7.1 Load fraction. The building service water-heating system shall have one or more of the following that are sized to provide not less than 60 percent of hot water requirements, or sized to provide 100 percent of hot water requirements if the building shall otherwise comply with Section C403.4.7:

1. Waste heat recovery from service hot water, heat-recovery chillers, building equipment, process equipment, or a combined heat and power system.
2. Solar water-heating systems.

SECTION C407
TOTAL BUILDING PERFORMANCE

C407.1 Scope. This section establishes criteria for compliance using total building performance. The following systems and loads shall be included in determining the total building performance: heating systems, cooling systems, service water heating, fan systems, lighting power, receptacle loads and process loads.

C407.2 Mandatory requirements. Compliance with this section requires that the criteria of Sections C402.5, C403.2, C404 and C405 be met.

C407.3 Performance-based compliance. Compliance based on total building performance requires that a proposed building (*proposed design*) be shown to have an annual energy cost that is less than or equal to the annual energy cost of the *standard reference design*. Energy prices shall be taken from a source *approved* by the *code official*, such as the Department of Energy, Energy Information Administration's *State Energy Price and Expenditure Report*. Code officials shall be permitted to require time-of-use pricing in energy cost calculations. Nondepletable energy collected off site shall be treated and priced the same as purchased energy. Energy from nondepletable energy sources collected on site shall be omitted from the annual energy cost of the *proposed design*.

Exception: Jurisdictions that require site energy (1 kWh = 3413 Btu) rather than energy cost as the metric of comparison.

C407.4 Documentation. Documentation verifying that the methods and accuracy of compliance software tools conform to the provisions of this section shall be provided to the *code official*.

C407.4.1 Compliance report. Permit submittals shall include a report documenting that the proposed design has annual energy costs less than or equal to the annual energy costs of the *standard reference design*. The compliance documentation shall include the following information:

1. Address of the building.

2. An inspection checklist documenting the building component characteristics of the *proposed design* as specified in Table C407.5.1(1). The inspection checklist shall show the estimated annual energy cost for both the *standard reference design* and the *proposed design*.

3. Name of individual completing the compliance report.

4. Name and version of the compliance software tool.

C407.4.2 Additional documentation. The *code official* shall be permitted to require the following documents:

1. Documentation of the building component characteristics of the *standard reference design*.

2. Thermal zoning diagrams consisting of floor plans showing the thermal zoning scheme for *standard reference design* and *proposed design*.

3. Input and output reports from the energy analysis simulation program containing the complete input and output files, as applicable. The output file shall include energy use totals and energy use by energy source and end-use served, total hours that space conditioning loads are not met and any errors or warning messages generated by the simulation tool as applicable.

4. An explanation of any error or warning messages appearing in the simulation tool output.

5. A certification signed by the builder providing the building component characteristics of the *proposed design* as given in Table C407.5.1(1).

C407.5 Calculation procedure. Except as specified by this section, the *standard reference design* and *proposed design* shall be configured and analyzed using identical methods and techniques.

C407.5.1 Building specifications. The *standard reference design* and *proposed design* shall be configured and analyzed as specified by Table C407.5.1(1). Table C407.5.1(1) shall include by reference all notes contained in Table C402.1.4.

C407.5.2 Thermal blocks. The *standard reference design* and *proposed design* shall be analyzed using identical thermal blocks as specified in Section C407.5.2.1, C407.5.2.2 or C407.5.2.3.

C407.5.2.1 HVAC zones designed. Where HVAC *zones* are defined on HVAC design drawings, each HVAC *zone* shall be modeled as a separate thermal block.

Exception: Different HVAC *zones* shall be allowed to be combined to create a single thermal block or identical thermal blocks to which multipliers are applied provided:

1. The space use classification is the same throughout the thermal block.

2. All HVAC *zones* in the thermal block that are adjacent to glazed exterior walls face the same orientation or their orientations are within 45 degrees (0.79 rad) of each other.

3. All of the *zones* are served by the same HVAC system or by the same kind of HVAC system.

C407.5.2.2 HVAC zones not designed. Where HVAC *zones* have not yet been designed, thermal blocks shall be defined based on similar internal load densities, occupancy, lighting, thermal and temperature schedules, and in combination with the following guidelines:

1. Separate thermal blocks shall be assumed for interior and perimeter spaces. Interior spaces shall be those located more than 15 feet (4572 mm) from an exterior wall. Perimeter spaces shall be those located closer than 15 feet (4572 mm) from an *exterior wall*.

2. Separate thermal blocks shall be assumed for spaces adjacent to glazed exterior walls: a separate *zone* shall be provided for each orientation, except orientations that differ by not more than 45 degrees (0.79 rad) shall be permitted to be considered to be the same orientation. Each *zone* shall include floor area that is 15 feet (4572 mm) or less from a glazed perimeter wall, except that floor area within 15 feet (4572 mm) of glazed perimeter walls having more than one orientation shall be divided proportionately between *zones*.

3. Separate thermal blocks shall be assumed for spaces having floors that are in contact with the ground or exposed to ambient conditions from *zones* that do not share these features.

4. Separate thermal blocks shall be assumed for spaces having exterior ceiling or roof assemblies from *zones* that do not share these features.

C407.5.2.3 Multifamily residential buildings. Residential spaces shall be modeled using one thermal block per space except that those facing the same orientations are permitted to be combined into one thermal block. Corner units and units with roof or floor loads shall only be combined with units sharing these features.

COMMERCIAL ENERGY EFFICIENCY

TABLE C407.5.1(1)
SPECIFICATIONS FOR THE STANDARD REFERENCE AND PROPOSED DESIGNS

BUILDING COMPONENT CHARACTERISTICS	STANDARD REFERENCE DESIGN	PROPOSED DESIGN
Space use classification	Same as proposed	The space use classification shall be chosen in accordance with Table C405.5.2 for all areas of the building covered by this permit. Where the space use classification for a building is not known, the building shall be categorized as an office building.
Roofs	Type: Insulation entirely above deck	As proposed
	Gross area: same as proposed	As proposed
	U-factor: as specified in Table C402.1.4	As proposed
	Solar absorptance: 0.75	As proposed
	Emittance: 0.90	As proposed
Walls, above-grade	Type: Mass wall where proposed wall is mass; otherwise steel-framed wall	As proposed
	Gross area: same as proposed	As proposed
	U-factor: as specified in Table C402.1.4	As proposed
	Solar absorptance: 0.75	As proposed
	Emittance: 0.90	As proposed
Walls, below-grade	Type: Mass wall	As proposed
	Gross area: same as proposed	As proposed
	U-Factor: as specified in Table C402.1.4 with insulation layer on interior side of walls	As proposed
Floors, above-grade	Type: joist/framed floor	As proposed
	Gross area: same as proposed	As proposed
	U-factor: as specified in Table C402.1.4	As proposed
Floors, slab-on-grade	Type: Unheated	As proposed
	F-factor: as specified in Table C402.1.4	As proposed
Opaque doors	Type: Swinging	As proposed
	Area: Same as proposed	As proposed
	U-factor: as specified in Table C402.1.4	As proposed
Vertical fenestration other than opaque doors	Area 1. The proposed glazing area; where the proposed glazing area is less than 40 percent of above-grade wall area. 2. 40 percent of above-grade wall area; where the proposed glazing area is 40 percent or more of the above-grade wall area.	As proposed
	U-factor: as specified in Table C402.4	As proposed
	SHGC: as specified in Table C402.4 except that for climates with no requirement (NR) SHGC = 0.40 shall be used	As proposed
	External shading and PF: None	As proposed
Skylights	Area 1. The proposed skylight area; where the proposed skylight area is less than 3 percent of gross area of roof assembly. 2. 3 percent of gross area of roof assembly; where the proposed skylight area is 3 percent or more of gross area of roof assembly	As proposed
	U-factor: as specified in Table C402.4	As proposed
	SHGC: as specified in Table C402.4 except that for climates with no requirement (NR) SHGC = 0.40 shall be used.	As proposed
Lighting, interior	The interior lighting power shall be determined in accordance with Section C405.4.2. Where the occupancy of the building is not known, the lighting power density shall be 1.0 Watt per square foot (10.7 W/m^2) based on the categorization of buildings with unknown space classification as offices.	As proposed
Lighting, exterior	The lighting power shall be determined in accordance with Table C405.5.2(2). Areas and dimensions of tradable and nontradable surfaces shall be the same as proposed.	As proposed

(continued)

TABLE C407.5.1(1)—continued
SPECIFICATIONS FOR THE STANDARD REFERENCE AND PROPOSED DESIGNS

BUILDING COMPONENT CHARACTERISTICS	STANDARD REFERENCE DESIGN	PROPOSED DESIGN
Internal gains	Same as proposed	Receptacle, motor and process loads shall be modeled and estimated based on the space use classification. All end-use load components within and associated with the building shall be modeled to include, but not be limited to, the following: exhaust fans, parking garage ventilation fans, exterior building lighting, swimming pool heaters and pumps, elevators, escalators, refrigeration equipment and cooking equipment.
Schedules	Same as proposed	Operating schedules shall include hourly profiles for daily operation and shall account for variations between weekdays, weekends, holidays and any seasonal operation. Schedules shall model the time-dependent variations in occupancy, illumination, receptacle loads, thermostat settings, mechanical ventilation, HVAC equipment availability, service hot water usage and any process loads. The schedules shall be typical of the proposed building type as determined by the designer and approved by the jurisdiction.
Mechanical ventilation	Same as proposed	As proposed, in accordance with Section C403.2.6.
Heating systems	Fuel type: same as proposed design	As proposed
	Equipment type[a]: as specified in Tables C407.5.1(2) and C407.5.1(3)	As proposed
	Efficiency: as specified in Tables C403.2.3(4) and C403.2.3(5)	As proposed
	Capacity[b]: sized proportionally to the capacities in the proposed design based on sizing runs, and shall be established such that no smaller number of unmet heating load hours and no larger heating capacity safety factors are provided than in the proposed design.	As proposed
Cooling systems	Fuel type: same as proposed design	As proposed
	Equipment type[c]: as specified in Tables C407.5.1(2) and C407.5.1(3)	As proposed
	Efficiency: as specified in Tables C403.2.3(1), C403.2.3(2) and C403.2.3(3)	As proposed
	Capacity[b]: sized proportionally to the capacities in the proposed design based on sizing runs, and shall be established such that no smaller number of unmet cooling load hours and no larger cooling capacity safety factors are provided than in the proposed design.	As proposed
	Economizer[d]: same as proposed, in accordance with Section C403.3.	As proposed
Service water heating[e]	Fuel type: same as proposed	As proposed
	Efficiency: as specified in Table C404.2	For Group R, as proposed multiplied by SWHF. For other than Group R, as proposed multiplied by efficiency as provided by the manufacturer of the DWHR unit.
	Capacity: same as proposed	As proposed
	Where no service water hot water system exists or is specified in the proposed design, no service hot water heating shall be modeled.	

SWHF = Service water heat recovery factor, DWHR = Drain water heat recovery.

a. Where no heating system exists or has been specified, the heating system shall be modeled as fossil fuel. The system characteristics shall be identical in both the standard reference design and proposed design.
b. The ratio between the capacities used in the annual simulations and the capacities determined by sizing runs shall be the same for both the standard reference design and proposed design.
c. Where no cooling system exists or no cooling system has been specified, the cooling system shall be modeled as an air-cooled single-zone system, one unit per thermal zone. The system characteristics shall be identical in both the standard reference design and proposed design.
d. If an economizer is required in accordance with Table C403.3 and where no economizer exists or is specified in the proposed design, then a supply-air economizer shall be provided in the standard reference design in accordance with Section C403.3.
e. The SWHF shall be applied as follows:
   1. Where potable water from the DWHR unit supplies not less than one shower and not greater than two showers, of which the drain water from the same showers flows through the DWHR unit then SWHF = [1 − (DWHR unit efficiency • 0.36)].
   2. Where potable water from the DWHR unit supplies not less than three showers and not greater than four showers, of which the drain water from the same showers flows through the DWHR unit then SWHF = [1 − (DWHR unit efficiency • 0.33)].
   3. Where potable water from the DWHR unit supplies not less than five showers and not greater than six showers, of which the drain water from the same showers flows through the DWHR unit, then SWHF = [1 − (DWHR unit efficiency • 0.26)].
   4. Where Items 1 through 3 are not met, SWHF = 1.0.

TABLE C407.5.1(2)
HVAC SYSTEMS MAP

CONDENSER COOLING SOURCE[a]	HEATING SYSTEM CLASSIFICATION[b]	STANDARD REFERENCE DESIGN HVC SYSTEM TYPE[c]		
		Single-zone Residential System	Single-zone Nonresidential System	All Other
Water/ground	Electric resistance	System 5	System 5	System 1
	Heat pump	System 6	System 6	System 6
	Fossil fuel	System 7	System 7	System 2
Air/none	Electric resistance	System 8	System 9	System 3
	Heat pump	System 8	System 9	System 3
	Fossil fuel	System 10	System 11	System 4

a. Select "water/ground" where the proposed design system condenser is water or evaporatively cooled; select "air/none" where the condenser is air cooled. Closed-circuit dry coolers shall be considered air cooled. Systems utilizing district cooling shall be treated as if the condenser water type were "water." Where no mechanical cooling is specified or the mechanical cooling system in the proposed design does not require heat rejection, the system shall be treated as if the condenser water type were "Air." For proposed designs with ground-source or groundwater-source heat pumps, the standard reference design HVAC system shall be water-source heat pump (System 6).

b. Select the path that corresponds to the proposed design heat source: electric resistance, heat pump (including air source and water source), or fuel fired. Systems utilizing district heating (steam or hot water) and systems with no heating capability shall be treated as if the heating system type were "fossil fuel." For systems with mixed fuel heating sources, the system or systems that use the secondary heating source type (the one with the smallest total installed output capacity for the spaces served by the system) shall be modeled identically in the standard reference design and the primary heating source type shall be used to determine *standard* reference design HVAC system type.

c. Select the standard reference design HVAC system category: The system under "single-zone residential system" shall be selected where the HVAC system in the proposed design is a single-zone system and serves a residential space. The system under "single-zone nonresidential system" shall be selected where the HVAC system in the proposed design is a single-zone system and serves other than residential spaces. The system under "all other" shall be selected for all other cases.

TABLE C407.5.1(3)
SPECIFICATIONS FOR THE STANDARD REFERENCE DESIGN HVAC SYSTEM DESCRIPTIONS

SYSTEM NO.	SYSTEM TYPE	FAN CONTROL	COOLING TYPE	HEATING TYPE
1	Variable air volume with parallel fan-powered boxes[a]	VAV[d]	Chilled water[e]	Electric resistance
2	Variable air volume with reheat[b]	VAV[d]	Chilled water[e]	Hot water fossil fuel boiler[f]
3	Packaged variable air volume with parallel fan-powered boxes[a]	VAV[d]	Direct expansion[c]	Electric resistance
4	Packaged variable air volume with reheat[b]	VAV[d]	Direct expansion[c]	Hot water fossil fuel boiler[f]
5	Two-pipe fan coil	Constant volume[i]	Chilled water[e]	Electric resistance
6	Water-source heat pump	Constant volume[i]	Direct expansion[c]	Electric heat pump and boiler[g]
7	Four-pipe fan coil	Constant volume[i]	Chilled water[e]	Hot water fossil fuel boiler[f]
8	Packaged terminal heat pump	Constant volume[i]	Direct expansion[c]	Electric heat pump[h]
9	Packaged rooftop heat pump	Constant volume[i]	Direct expansion[c]	Electric heat pump[h]
10	Packaged terminal air conditioner	Constant volume[i]	Direct expansion	Hot water fossil fuel boiler[f]
11	Packaged rooftop air conditioner	Constant volume[i]	Direct expansion	Fossil fuel furnace

For SI: 1 foot = 304.8 mm, 1 cfm/ft² = 0.4719 L/s, 1 Btu/h = 0.293/W, °C = [(°F) -32/1.8].

a. **VAV with parallel boxes:** Fans in parallel VAV fan-powered boxes shall be sized for 50 percent of the peak design flow rate and shall be modeled with 0.35 W/cfm fan power. Minimum volume setpoints for fan-powered boxes shall be equal to the minimum rate for the space required for ventilation consistent with Section C403.4.4, Exception 4. Supply air temperature setpoint shall be constant at the design condition.

b. **VAV with reheat:** Minimum volume setpoints for VAV reheat boxes shall be 0.4 cfm/ft² of floor area. Supply air temperature shall be reset based on zone demand from the design temperature difference to a 10°F temperature difference under minimum load conditions. Design airflow rates shall be sized for the reset supply air temperature, i.e., a 10°F temperature difference.

c. **Direct expansion:** The fuel type for the cooling system shall match that of the cooling system in the proposed design.

d. **VAV:** Where the proposed design system has a supply, return or relief fan motor 25 hp or larger, the corresponding fan in the VAV system of the standard reference design shall be modeled assuming a variable-speed drive. For smaller fans, a forward-curved centrifugal fan with inlet vanes shall be modeled. Where the proposed design's system has a direct digital control system at the zone level, static pressure setpoint reset based on zone requirements in accordance with Section C403.4.1 shall be modeled.

e. **Chilled water:** For systems using purchased chilled water, the chillers are not explicitly modeled and chilled water costs shall be based as determined in Sections C407.3 and C407.5.2. Otherwise, the standard reference design's chiller plant shall be modeled with chillers having the number as indicated in Table C407.5.1(4) as a function of standard reference building chiller plant load and type as indicated in Table C407.5.1(5) as a function of individual chiller load. Where chiller fuel source is mixed, the system in the standard reference design shall have chillers with the same fuel types and with capacities having the same proportional capacity as the proposed design's chillers for each fuel type. Chilled water supply temperature shall be modeled at 44°F design supply temperature and 56°F return temperature. Piping losses shall not be modeled in either building model. Chilled water supply water temperature shall be reset in accordance with Section C403.4.3.3. Pump system power for each pumping system shall be the same as the proposed design; where the proposed design has no chilled water pumps, the standard reference design pump power shall be 22 W/gpm (equal to a pump operating against a 75-foot head, 65-percent combined impeller and motor efficiency). The chilled water system shall be modeled as primary-only variable flow with flow maintained at the design rate through each chiller using a bypass. Chilled water pumps shall be modeled as riding the pump curve or with variable-speed drives when required in Section C403.4.3.3. The heat rejection device shall be an axial fan cooling tower with two-speed fans where required in Section C403.4.3. Condenser water design supply temperature shall be 85°F or 10°F approach to design wet-bulb temperature, whichever is lower, with a design temperature rise of 10°F. The tower shall be controlled to maintain a 70°F leaving water temperature where weather permits, floating up to leaving water temperature at design conditions. Pump system power for each pumping system shall be the same as the proposed design; where the proposed design has no condenser water pumps, the standard reference design pump power shall be 19 W/gpm (equal to a pump operating against a 60-foot head, 60-percent combined impeller and motor efficiency). Each chiller shall be modeled with separate condenser water and chilled water pumps interlocked to operate with the associated chiller.

f. **Fossil fuel boiler:** For systems using purchased hot water or steam, the boilers are not explicitly modeled and hot water or steam costs shall be based on actual utility rates. Otherwise, the boiler plant shall use the same fuel as the proposed design and shall be natural draft. The standard reference design boiler plant shall be modeled with a single boiler where the standard reference design plant load is 600,000 Btu/h and less and with two equally sized boilers for plant capacities exceeding 600,000 Btu/h. Boilers shall be staged as required by the load. Hot water supply temperature shall be modeled at 180°F design supply temperature and 130°F return temperature. Piping losses shall not be modeled in either building model. Hot water supply water temperature shall be reset in accordance with Section C403.4.3.3. Pump system power for each pumping system shall be the same as the proposed design; where the proposed design has no hot water pumps, the standard reference design pump power shall be 19 W/gpm (equal to a pump operating against a 60-foot head, 60-percent combined impeller and motor efficiency). The hot water system shall be modeled as primary only with continuous variable flow. Hot water pumps shall be modeled as riding the pump curve or with variable speed drives when required by Section C403.4.3.3.

g. **Electric heat pump and boiler:** Water-source heat pumps shall be connected to a common heat pump water loop controlled to maintain temperatures between 60°F and 90°F. Heat rejection from the loop shall be provided by an axial fan closed-circuit evaporative fluid cooler with two-speed fans where required in Section C403.4.1. Heat addition to the loop shall be provided by a boiler that uses the same fuel as the proposed design and shall be natural draft. Where no boilers exist in the proposed design, the standard reference building boilers shall be fossil fuel. The standard reference design boiler plant shall be modeled with a single boiler where the standard reference design plant load is 600,000 Btu/h or less and with two equally sized boilers for plant capacities exceeding 600,000 Btu/h. Boilers shall be staged as required by the load. Piping losses shall not be modeled in either building model. Pump system power shall be the same as the proposed design; where the proposed design has no pumps, the standard reference design pump power shall be 22 W/gpm, which is equal to a pump operating against a 75-foot head, with a 65-percent combined impeller and motor efficiency. Loop flow shall be variable with flow shutoff at each heat pump when its compressor cycles off as required by Section C403.4.3.3. Loop pumps shall be modeled as riding the pump curve or with variable speed drives when required by Section C403.4.3.3.

h. **Electric heat pump:** Electric air-source heat pumps shall be modeled with electric auxiliary heat. The system shall be controlled with a multistage space thermostat and an outdoor air thermostat wired to energize auxiliary heat only on the last thermostat stage and when outdoor air temperature is less than 40°F.

i. **Constant volume:** Fans shall be controlled in the same manner as in the proposed design; i.e., fan operation whenever the space is occupied or fan operation cycled on calls for heating and cooling. Where the fan is modeled as cycling and the fan energy is included in the energy efficiency rating of the equipment, fan energy shall not be modeled explicitly.

TABLE C407.5.1(4)
NUMBER OF CHILLERS

TOTAL CHILLER PLANT CAPACITY	NUMBER OF CHILLERS
≤ 300 tons	1
> 300 tons, < 600 tons	2, sized equally
≥ 600 tons	2 minimum, with chillers added so that no chiller is larger than 800 tons, all sized equally

For SI: 1 ton = 3517 W.

TABLE C407.5.1(5)
WATER CHILLER TYPES

INDIVIDUAL CHILLER PLANT CAPACITY	ELECTRIC CHILLER TYPE	FOSSIL FUEL CHILLER TYPE
≤ 100 tons	Reciprocating	Single-effect absorption, direct fired
> 100 tons, < 300 tons	Screw	Double-effect absorption, direct fired
≥ 300 tons	Centrifugal	Double-effect absorption, direct fired

For SI: 1 ton = 3517 W.

C407.6 Calculation software tools. Calculation procedures used to comply with this section shall be software tools capable of calculating the annual energy consumption of all building elements that differ between the *standard reference design* and the *proposed design* and shall include the following capabilities.

1. Building operation for a full calendar year (8,760 hours).
2. Climate data for a full calendar year (8,760 hours) and shall reflect *approved* coincident hourly data for temperature, solar radiation, humidity and wind speed for the building location.
3. Ten or more thermal zones.
4. Thermal mass effects.
5. Hourly variations in occupancy, illumination, receptacle loads, thermostat settings, mechanical ventilation, HVAC equipment availability, service hot water usage and any process loads.
6. Part-load performance curves for mechanical equipment.
7. Capacity and efficiency correction curves for mechanical heating and cooling equipment.
8. Printed *code official* inspection checklist listing each of the *proposed design* component characteristics from Table C407.5.1(1) determined by the analysis to provide compliance, along with their respective performance ratings including, but not limited to, *R*-value, *U* factor, SHGC, HSPF, AFUE, SEER, EF.

C407.6.1 Specific approval. Performance analysis tools complying with the applicable subsections of Section C407 and tested according to ASHRAE Standard 140 shall be permitted to be *approved*. Tools are permitted to be *approved* based on meeting a specified threshold for a jurisdiction. The *code official* shall be permitted to approve tools for a specified application or limited scope.

C407.6.2 Input values. Where calculations require input values not specified by Sections C402, C403, C404 and C405, those input values shall be taken from an *approved* source.

C407.6.3 Exceptional calculation methods. Where the simulation program does not model a design, material or device of the *proposed design*, an exceptional calculation method shall be used where approved by the *code official*. Where there are multiple designs, materials or devices that the simulation program does not model, each shall be calculated separately and exceptional savings determined for each. The total exceptional savings shall not constitute more than half of the difference between the baseline building performance and the proposed building performance. Applications for approval of an exceptional method shall include all of the following:

1. Step-by-step documentation of the exceptional calculation method performed, detailed enough to reproduce the results.
2. Copies of all spreadsheets used to perform the calculations.
3. A sensitivity analysis of energy consumption where each of the input parameters is varied from half to double the value assumed.
4. The calculations shall be performed on a time step basis consistent with the simulation program used.
5. The performance rating calculated with and without the exceptional calculation method.

SECTION C408
SYSTEM COMMISSIONING

C408.1 General. This section covers the commissioning of the building mechanical systems in Section C403 and electrical power and lighting systems in Section C405.

C408.2 Mechanical systems and service water-heating systems commissioning and completion requirements. Prior to the final mechanical and plumbing inspections, the *registered design professional* or *approved agency* shall provide evidence of mechanical systems *commissioning* and completion in accordance with the provisions of this section.

Construction document notes shall clearly indicate provisions for *commissioning* and completion requirements in accordance with this section and are permitted to refer to specifications for further requirements. Copies of all documentation shall be given to the owner or owner's authorized agent and made available to the *code official* upon request in accordance with Sections C408.2.4 and C408.2.5.

Exceptions: The following systems are exempt:

1. Mechanical systems and service water heater systems in buildings where the total mechanical equipment capacity is less than 480,000 Btu/h (140.7 kW) cooling capacity and 600,000 Btu/h (175.8 kW) combined service water-heating and space-heating capacity.
2. Systems included in Section C403.3 that serve individual *dwelling units* and *sleeping units*.

C408.2.1 Commissioning plan. A *commissioning plan* shall be developed by a *registered design professiona*l or *approved agency* and shall include the following items:

1. A narrative description of the activities that will be accomplished during each phase of *commissioning,* including the personnel intended to accomplish each of the activities.
2. A listing of the specific equipment, appliances or systems to be tested and a description of the tests to be performed.
3. Functions to be tested including, but not limited to, calibrations and economizer controls.
4. Conditions under which the test will be performed. Testing shall affirm winter and summer design conditions and full outside air conditions.
5. Measurable criteria for performance.

C408.2.2 Systems adjusting and balancing. HVAC systems shall be balanced in accordance with generally accepted engineering standards. Air and water flow rates shall be measured and adjusted to deliver final flow rates within the tolerances provided in the product specifications. Test and balance activities shall include air system and hydronic system balancing.

C408.2.2.1 Air systems balancing. Each supply air outlet and *zone* terminal device shall be equipped with means for air balancing in accordance with the requirements of Chapter 6 of the *International Mechanical Code*. Discharge dampers used for air-system balancing are prohibited on constant-volume fans and variable-volume fans with motors 10 hp (18.6 kW) and larger. Air systems shall be balanced in a manner to first minimize throttling losses then, for fans with system power of greater than 1 hp (0.746 kW), fan speed shall be adjusted to meet design flow conditions.

Exception: Fans with fan motors of 1 hp (0.74 kW) or less are not required to be provided with a means for air balancing.

C408.2.2.2 Hydronic systems balancing. Individual hydronic heating and cooling coils shall be equipped with means for balancing and measuring flow. Hydronic systems shall be proportionately balanced in a manner to first minimize throttling losses, then the pump impeller shall be trimmed or pump speed shall be adjusted to meet design flow conditions. Each hydronic system shall have either the capability to measure pressure across the pump, or test ports at each side of each pump.

Exceptions: The following equipment is not required to be equipped with a means for balancing or measuring flow:

1. Pumps with pump motors of 5 hp (3.7 kW) or less.
2. Where throttling results in no greater than 5 percent of the nameplate horsepower draw above that required if the impeller were trimmed.

C408.2.3 Functional performance testing. Functional performance testing specified in Sections C408.2.3.1 through C408.2.3.3 shall be conducted.

C408.2.3.1 Equipment. Equipment functional performance testing shall demonstrate the installation and operation of components, systems, and system-to-system interfacing relationships in accordance with approved plans and specifications such that operation, function, and maintenance serviceability for each of the commissioned systems is confirmed. Testing shall include all modes and *sequence of operation*, including under full-load, part-load and the following emergency conditions:

1. All modes as described in the *sequence* of *operation.*
2. Redundant or *automatic* back-up mode.
3. Performance of alarms.
4. Mode of operation upon a loss of power and restoration of power.

Exception: Unitary or packaged HVAC equipment listed in Tables C403.2.3(1) through C403.2.3(3) that do not require supply air economizers.

C408.2.3.2 Controls. HVAC and service water-heating control systems shall be tested to document that control devices, components, equipment and systems are calibrated and adjusted and operate in accordance with approved plans and specifications. Sequences of operation shall be functionally tested to document they operate in accordance with *approved* plans and specifications.

C408.2.3.3 Economizers. Air economizers shall undergo a functional test to determine that they operate in accordance with manufacturer's specifications.

C408.2.4 Preliminary commissioning report. A preliminary report of *commissioning* test procedures and results shall be completed and certified by the *registered design professional* or *approved agency* and provided to the building owner or owner's authorized agent. The report shall be organized with mechanical and service hot water findings in separate sections to allow independent review. The report shall be identified as "Preliminary Commissioning Report" and shall identify:

1. Itemization of deficiencies found during testing required by this section that have not been corrected at the time of report preparation.
2. Deferred tests that cannot be performed at the time of report preparation because of climatic conditions.
3. Climatic conditions required for performance of the deferred tests.

C408.2.4.1 Acceptance of report. Buildings, or portions thereof, shall not be considered acceptable for a final inspection pursuant to Section C104.3 until the *code official* has received a letter of transmittal from the building owner acknowledging that the building owner or owner's authorized agent has received the Preliminary Commissioning Report.

C408.2.4.2 Copy of report. The *code official* shall be permitted to require that a copy of the Preliminary Commissioning Report be made available for review by the *code official*.

C408.2.5 Documentation requirements. The *construction documents* shall specify that the documents described in this section be provided to the building owner or owner's authorized agent within 90 days of the date of receipt of the *certificate of occupancy*.

C408.2.5.1 Drawings. *Construction documents* shall include the location and performance data on each piece of equipment.

C408.2.5.2 Manuals. An operating and maintenance manual shall be provided and include all of the following:

1. Submittal data stating equipment size and selected options for each piece of equipment requiring maintenance.
2. Manufacturer's operation manuals and maintenance manuals for each piece of equipment requiring maintenance, except equipment not furnished as part of the project. Required routine maintenance actions shall be clearly identified.
3. Name and address of at least one service agency.
4. HVAC and service hot water controls system maintenance and calibration information, including wiring diagrams, schematics and control sequence descriptions. Desired or field-determined set points shall be permanently recorded on control drawings at control devices or, for digital control systems, in system programming instructions.
5. Submittal data indicating all selected options for each piece of lighting equipment and lighting controls.
6. Operation and maintenance manuals for each piece of lighting equipment. Required routine maintenance actions, cleaning and recommended relamping shall be clearly identified.
7. A schedule for inspecting and recalibrating all lighting controls.
8. A narrative of how each system is intended to operate, including recommended set points.

C408.2.5.3 System balancing report. A written report describing the activities and measurements completed in accordance with Section C408.2.2.

C408.2.5.4 Final commissioning report. A report of test procedures and results identified as "Final Commissioning Report" shall be delivered to the building owner or owner's authorized agent. The report shall be organized with mechanical system and service hot water system findings in separate sections to allow independent review. The report shall include the following:

1. Results of functional performance tests.
2. Disposition of deficiencies found during testing, including details of corrective measures used or proposed.
3. Functional performance test procedures used during the commissioning process including measurable criteria for test acceptance, provided herein for repeatability.

Exception: Deferred tests that cannot be performed at the time of report preparation due to climatic conditions.

C408.3 Lighting system functional testing. Controls for automatic lighting systems shall comply with this section.

C408.3.1 Functional testing. Prior to passing final inspection, the *registered design professional* shall provide evidence that the lighting control systems have been tested to ensure that control hardware and software are calibrated, adjusted, programmed and in proper working condition in accordance with the *construction documents* and manufacturer's instructions. Functional testing shall be in accordance with Sections C408.3.1.1 and C408.3.1.2 for the applicable control type.

C408.3.1.1 Occupant sensor controls. Where *occupant sensor controls* are provided, the following procedures shall be performed:

1. Certify that the *occupant sensor* has been located and aimed in accordance with manufacturer recommendations.
2. For projects with seven or fewer *occupant sensors,* each sensor shall be tested.

3. For projects with more than seven *occupant sensors*, testing shall be done for each unique combination of sensor type and space geometry. Where multiples of each unique combination of sensor type and space geometry are provided, not less than 10 percent, but in no case less than one, of each combination shall be tested unless the *code official* or design professional requires a higher percentage to be tested. Where 30 percent or more of the tested controls fail, all remaining identical combinations shall be tested.

   For *occupant sensor controls* to be tested, verify the following:

   3.1. Where *occupant sensor controls* include status indicators, verify correct operation.

   3.2. The controlled lights turn off or down to the permitted level within the required time.

   3.3. For auto-on *occupant sensor controls*, the lights turn on to the permitted level when an occupant enters the space.

   3.4. For manual-on *occupant sensor controls*, the lights turn on only when manually activated.

   3.5. The lights are not incorrectly turned on by movement in adjacent areas or by HVAC operation.

C408.3.1.2 Time-switch controls. Where t*ime-switch controls* are provided, the following procedures shall be performed:

1. Confirm that the *time-switch control* is programmed with accurate weekday, weekend and holiday schedules.

2. Provide documentation to the owner of *time-switch controls* programming including weekday, weekend, holiday schedules, and set-up and preference program settings.

3. Verify the correct time and date in the time switch.

4. Verify that any battery back-up is installed and energized.

5. Verify that the override time limit is set to not more than 2 hours.

6. Simulate occupied condition. Verify and document the following:

   6.1. All lights can be turned on and off by their respective area control switch.

   6.2. The switch only operates lighting in the enclosed space in which the switch is located.

7. Simulate unoccupied condition. Verify and document the following:

   7.1. Nonexempt lighting turns off.

   7.2. Manual override switch allows only the lights in the enclosed space where the override switch is located to turn on or remain on until the next scheduled shutoff occurs.

8. Additional testing as specified by the *registered design professional*.

C408.3.1.3 Daylight responsive controls. Where *daylight responsive controls* are provided, the following shall be verified:

1. Control devices have been properly located, field calibrated and set for accurate setpoints and threshold light levels.

2. Daylight controlled lighting loads adjust to light level set points in response to available daylight.

3. The locations of calibration adjustment equipment are readily accessible only to authorized personnel.

C408.3.2 Documentation requirements. The *construction documents* shall specify that documents certifying that the installed lighting controls meet documented performance criteria of Section C405 are to be provided to the building owner within 90 days from the date of receipt of the *certificate of occupancy*.

CHAPTER 5 [CE]
EXISTING BUILDINGS

SECTION C501
GENERAL

C501.1 Scope. The provisions of this chapter shall control the *alteration*, *repair*, *addition* and change of occupancy of existing buildings and structures.

C501.2 Existing buildings. Except as specified in this chapter, this code shall not be used to require the removal, *alteration* or abandonment of, nor prevent the continued use and maintenance of, an existing building or building system lawfully in existence at the time of adoption of this code.

C501.3 Maintenance. Buildings and structures, and parts thereof, shall be maintained in a safe and sanitary condition. Devices and systems that are required by this code shall be maintained in conformance to the code edition under which installed. The owner or the owner's authorized agent shall be responsible for the maintenance of buildings and structures. The requirements of this chapter shall not provide the basis for removal or abrogation of energy conservation, fire protection and safety systems and devices in existing structures.

C501.4 Compliance. *Alterations*, *repairs*, *additions* and changes of occupancy to, or relocation of, existing buildings and structures shall comply with the provisions for *alterations*, *repairs*, *additions* and changes of occupancy or relocation, respectively, in the *International Building Code*, *International Fire Code*, *International Fuel Gas Code*, *International Mechanical Code*, *International Plumbing Code*, *International Property Maintenance Code*, *International Private Sewage Disposal Code* and NFPA 70.

C501.5 New and replacement materials. Except as otherwise required or permitted by this code, materials permitted by the applicable code for new construction shall be used. Like materials shall be permitted for *repairs*, provided hazards to life, health or property are not created. Hazardous materials shall not be used where the code for new construction would not permit use of these materials in buildings of similar occupancy, purpose and location.

C501.6 Historic buildings. No provisions of this code relating to the construction, *repair, alteration*, restoration and movement of structures, and *change of occupancy* shall be mandatory for *historic buildings* provided a report has been submitted to the *code official* and signed by a *registered design professional*, or a representative of the State Historic Preservation Office or the historic preservation authority having jurisdiction, demonstrating that compliance with that provision would threaten, degrade or destroy the historic form, fabric or function of the building.

SECTION C502
ADDITIONS

C502.1 General. *Additions* to an existing building, building system or portion thereof shall conform to the provisions of this code as those provisions relate to new construction without requiring the unaltered portion of the existing building or building system to comply with this code. *Additions* shall not create an unsafe or hazardous condition or overload existing building systems. An *addition* shall be deemed to comply with this code if the *addition* alone complies or if the existing building and *addition* comply with this code as a single building. *Additions* shall comply with Section C502.2.

Additions complying with ANSI/ASHRAE/IESNA 90.1. need not comply with Sections C402, C403, C404 and C405.

C502.2 Prescriptive compliance. *Additions* shall comply with Sections C502.2.1 through C502.2.6.2.

C502.2.1 Vertical fenestration. New *vertical fenestration* area that results in a total building *fenestration* area less than or equal to that specified in Section C402.4.1 shall comply with Section C402.4. *Additions* with *vertical fenestration* that result in a total building *fenestration* area greater than Section C402.4.1 or *additions* that exceed the fenestration area greater than Section C402.4.1 shall comply with Section C402.4.1.1 for the *addition* only. *Additions* that result in a total building vertical glass area exceeding that specified in Section C402.4.1.1 shall comply with Section C407.

C502.2.2 Skylight area. New *skylight* area that results in a total building *fenestration* area less than or equal to that specified in Section C402.4.1 shall comply with Section C402.4. *Additions* with *skylight* area that result in a total building *skylight* area greater than C402.4.1 or additions that exceed the *skylight* area shall comply with Section C402.4.1.2 for the *addition* only. *Additions* that result in a total building *skylight* area exceeding that specified in Section C402.4.1.2 shall comply with Section C407.

C502.2.3 Building mechanical systems. New mechanical systems and equipment that are part of the *addition* and serve the building heating, cooling and ventilation needs shall comply with Section C403.

C502.2.4 Service water-heating systems. New service water-heating equipment, controls and service water heating piping shall comply with Section C404.

C502.2.5 Pools and inground permanently installed spas. New pools and inground permanently installed spas shall comply with Section C404.9.

C502.2.6 Lighting power and systems. New lighting systems that are installed as part of the addition shall comply with Section C405.

C502.2.6.1 Interior lighting power. The total interior lighting power for the *addition* shall comply with Section C405.4.2 for the *addition* alone, or the existing building and the *addition* shall comply as a single building.

C502.2.6.2 Exterior lighting power. The total exterior lighting power for the *addition* shall comply with Section C405.5.1 for the *addition* alone, or the existing building and the *addition* shall comply as a single building.

SECTION C503
ALTERATIONS

C503.1 General. *Alterations* to any building or structure shall comply with the requirements of the code for new construction. *Alterations* shall be such that the existing building or structure is no less conforming to the provisions of this code than the existing building or structure was prior to the *alteration*. *Alterations* to an existing building, building system or portion thereof shall conform to the provisions of this code as those provisions relate to new construction without requiring the unaltered portions of the existing building or building system to comply with this code. *Alterations* shall not create an unsafe or hazardous condition or overload existing building systems.

Alterations complying with ANSI/ASHRAE/IESNA 90.1. need not comply with Sections C402, C403, C404 and C405.

> **Exception:** The following *alterations* need not comply with the requirements for new construction, provided the energy use of the building is not increased:
>
> 1. Storm windows installed over existing *fenestration*.
> 2. Surface-applied window film installed on existing single-pane *fenestration* assemblies reducing solar heat gain, provided the code does not require the glazing or *fenestration* to be replaced.
> 3. Existing ceiling, wall or floor cavities exposed during construction, provided that these cavities are filled with insulation.
> 4. Construction where the existing roof, wall or floor cavity is not exposed.
> 5. *Roof recover.*
> 6. *Air barriers* shall not be required for *roof recover* and roof replacement where the *alterations* or renovations to the building do not include *alterations*, renovations or *repairs* to the remainder of the building envelope.
> 7. *Alterations* that replace less than 50 percent of the luminaires in a space, provided that such *alterations* do not increase the installed interior lighting power.

C503.2 Change in space conditioning. Any nonconditioned or low-energy space that is altered to become conditioned space shall be required to be brought into full compliance with this code.

C503.3 Building envelope. New building envelope assemblies that are part of the *alteration* shall comply with Sections C402.1 through C402.5.

C503.3.1 Roof replacement. *Roof replacements* shall comply with Table C402.1.3 or C402.1.4 where the existing roof assembly is part of the *building thermal envelope* and contains insulation entirely above the roof deck.

C503.3.2 Vertical fenestration. The addition of *vertical fenestration* that results in a total building *fenestration* area less than or equal to that specified in Section C402.4.1 shall comply with Section C402.4. The addition of *vertical fenestration* that results in a total building *fenestration* area greater than Section C402.4.1 shall comply with Section C402.4.1.1 for the space adjacent to the new fenestration only. *Alterations* that result in a total building vertical glass area exceeding that specified in Section C402.4.1.1 shall comply with Section C407.

C503.3.3 Skylight area. The addition of *skylight* area that results in a total building *skylight* area less than or equal to that specified in Section C402.4.1 shall comply with Section C402.4. The addition of *skylight* area that results in a total building skylight area greater than Section C402.4.1 shall comply with Section C402.4.1.2 for the space adjacent to the new skylights. *Alterations* that result in a total building skylight area exceeding that specified in Section C402.4.1.2 shall comply with Section C407.

C503.4 Heating and cooling systems. New heating, cooling and duct systems that are part of the *alteration* shall comply with Sections C403.

C503.4.1 Economizers. New cooling systems that are part of *alteration* shall comply with Section C403.3.

C503.5 Service hot water systems. New service hot water systems that are part of the *alteration* shall comply with Section C404.

C503.6 Lighting systems. New lighting systems that are part of the *alteration* shall comply with Section C405.

> **Exception.** *Alterations* that replace less than 10 percent of the luminaires in a space, provided that such *alterations* do not increase the installed interior lighting power.

SECTION C504
REPAIRS

C504.1 General. Buildings and structures, and parts thereof, shall be repaired in compliance with Section C501.3 and this section. Work on nondamaged components that is necessary for the required *repair* of damaged components shall be considered part of the *repair* and shall not be subject to the requirements for *alterations* in this chapter. Routine maintenance required by Section C501.3, ordinary *repairs* exempt from *permit* and abatement of wear due to normal service conditions shall not be subject to the requirements for *repairs* in this section.

Where a building was constructed to comply with ANSI/ASHRAE/IESNA 90.1, repairs shall comply with the standard and need not comply with Sections C402, C403, C404 and C405.

C504.2 Application. For the purposes of this code, the following shall be considered repairs:

1. Glass-only replacements in an existing sash and frame.

2. *Roof repairs.*

3. Air barriers shall not be required for *roof repair* where the repairs to the building do not include *alterations*, renovations or *repairs* to the remainder of the building envelope.

4. Replacement of existing doors that separate conditioned space from the exterior shall not require the installation of a vestibule or revolving door, provided that an existing vestibule that separates a conditioned space from the exterior shall not be removed.

5. *Repairs* where only the bulb, the ballast or both within the existing luminaires in a space are replaced, provided that the replacement does not increase the installed interior lighting power.

SECTION C505
CHANGE OF OCCUPANCY OR USE

C505.1 General. Spaces undergoing a change in occupancy that would result in an increase in demand for either fossil fuel or electrical energy shall comply with this code. Where the use in a space changes from one use in Table C405.4.2(1) or C405.4.2(2) to another use in Table C405.4.2(1) or C405.4.2(2), the installed lighting wattage shall comply with Section C405.4.

CHAPTER 6 [CE]

REFERENCED STANDARDS

This chapter lists the standards that are referenced in various sections of this document. The standards are listed herein by the promulgating agency of the standard, the standard identification, the effective date and title, and the section or sections of this document that reference the standard. The application of the referenced standards shall be as specified in Section 106.

AAMA

American Architectural Manufacturers Association
1827 Walden Office Square
Suite 550
Schaumburg, IL 60173-4268

Standard reference number	Title	Referenced in code section number
AAMA/WDMA/CSA 101/I.S.2/A C440—11	North American Fenestration Standard/ Specifications for Windows, Doors and Unit Skylights	Table C402.5.2

AHAM

Association of Home Appliance Manufacturers
1111 19th Street, NW, Suite 402
Washington, DC 20036

Standard reference number	Title	Referenced in code section number
ANSI/ AHAM RAC-1—2008	Room Air Conditioners	Table C403.2.3(3)
AHAM HRF-1—2007	Energy, Performance and Capacity of Household Refrigerators, Refrigerator-Freezers and Freezers	Table C403.2.14.1

AHRI

Air-Conditioning, Heating, and Refrigeration Institute
2111 Wilson Blvd, Suite 500
Arlington, VA 22201

Standard reference number	Title	Referenced in code section number
ISO/AHRI/ASHRAE 13256-1 (2011)	Water-to-Air and Brine-to-Air Heat Pumps—Testing and Rating for Performance	Table C403.2.3(2)
ISO/AHRI/ASHRAE 13256-2 (2011)	Water-to-Water and Brine-to-Water Heat Pumps —Testing and Rating for Performance	Table C403.2.3(2)
210/240—08 with Addenda 1 and 2	Performance Rating of Unitary Air-Conditioning and Air-Source Heat Pump Equipment	Table C403.2.3(1), Table C403.2.3(2)
310/380—04 (CSA-C744-04)	Standard for Packaged Terminal Air Conditioners and Heat Pumps	Table C403.2.3(3)
340/360—2007 with Addendum 2	Performance Rating of Commercial and Industrial Unitary Air-Conditioning and Heat Pump Equipment	Table C403.2.3(1), Table C403.2.3(2)
365(I-P)—09	Commercial and Industrial Unitary Air-Conditioning Condensing Units	Table C403.2.3(1), Table C403.2.3(6)
390—03	Performance Rating of Single Package Vertical Air-Conditioners and Heat Pumps	Table C403.2.3(3)
400—2001	Liquid to Liquid Heat Exchangers with Addendum 1 and 2	Table C403.2.3(10)
440—2008	Performance Rating of Room Fan Coils	C403.2.10
460—2005	Performance Rating of Remote Mechanical-Draft Air-Cooled Refrigerant Condensers	Table C403.2.3(8)

REFERENCED STANDARDS

AHRI—continued

550/590—2011 With Addendum 1	Performance Rating of Water-Chilling and Heat Pump Water-Heating Packages Using the Vapor Compression Cycle	C403.2.3.1, Table C403.2.3(7)
560—00	Absorption Water Chilling and Water Heating Packages	Table C403.2.3(7)
1160 (I-P) —09	Performance Rating of Heat Pump Pool Heaters	Table C404.2
1200-2010	Performance Rating of Commercial Refrigerated Display Merchandisers and Storage Cabinets	C403.2.14, Table C403.2.14(1), Table C403.2.14(2)

AMCA

Air Movement and Control Association International
30 West University Drive
Arlington Heights, IL 60004-1806

Standard reference number	Title	Referenced in code section number
205—12	Energy Efficiency Classification for Fans	C403.2.12.3
220—08 (R2012)	Laboratory Methods of Testing Air Curtain Units for Aerodynamic Performance Rating	C402.5.6
500D—12	Laboratory Methods for Testing Dampers for Rating	C403.2.4.3

ANSI

American National Standards Institute
25 West 43rd Street
Fourth Floor
New York, NY 10036

Standard reference number	Title	Referenced in code section number
Z21.10.3/CSA 4.3—11	Gas Water Heaters, Volume III—Storage Water Heaters with Input Ratings Above 75,000 Btu per Hour, Circulating Tank and Instantaneous	Table C404.2
Z21.47/CSA 2.3—12	Gas-fired Central Furnaces	Table C403.2.3(4)
Z83.8/CSA 2.6—09	Gas Unit Heaters, Gas Packaged Heaters, Gas Utility Heaters and Gas-fired Duct Furnaces	Table C403.2.3(4)

APSP

The Association of Pool & Spa Professionals
2111 Eisenhower Avenue
Alexandria, VA 22314

Standard reference number	Title	Referenced in code section number
14—11	American National Standard for Portable Electric Spa Efficiency	C404.8

ASHRAE

American Society of Heating, Refrigerating and Air-Conditioning Engineers, Inc.
1791 Tullie Circle, NE
Atlanta, GA 30329-2305

Standard reference number	Title	Referenced in code section number
ASHRAE 127-2007	Method of Testing for Rating Computer	Table C403.2.3(9)
ANSI/ASHRAE/ACCA Standard 183—2007 (RA2011)	Peak Cooling and Heating Load Calculations in Buildings, Except Low-rise Residential Buildings	C403.2.1
ASHRAE—2012	ASHRAE HVAC Systems and Equipment Handbook	C403.2.1
ISO/AHRI/ASHRAE 13256-1 (2011)	Water-to-Air and Brine-to-Air Heat Pumps—Testing and Rating for Performance	Table C403.2.3(2)

REFERENCED STANDARDS

ASHRAE—continued

ISO/AHRI/ASHRAE		
13256-2 (2011)	Water-to-Water and Brine-to-Water Heat Pumps—Testing and Rating for Performance	Table C403.2.3(2)
90.1—2013	Energy Standard for Buildings Except Low-rise Residential Buildings	C401.2, Table C402.1.3, Table C402.1.4, C406.2, Table C407.6.1, C502.1, C503.1, C504.1
140—2011	Standard Method of Test for the Evaluation of Building Energy Analysis Computer Programs	C407.6.1
146—2011	Testing and Rating Pool Heaters	Table C404.2

ASME

American Society Mechanical Engineers
Two Park Avenue
New York, NY 10016-5990

Standard reference number	Title	Referenced in code section number
ASME A17.1/ CSA B44—2013	Safety Code for Elevators and Escalators	C405.9.2

ASTM

ASTM International
100 Barr Harbor Drive
West Conshohocken, PA 19428-2859

Standard reference number	Title	Referenced in code section number
C 90—13	Specification for Load-bearing Concrete Masonry Units	Table C401.3
C 1363—11	Standard Test Method for Thermal Performance of Building Materials and Envelope Assemblies by Means of a Hot Box Apparatus	C303.1.4.1, Table C402.1.4
C 1371—04a(2010)e1	Standard Test Method for Determination of Emittance of Materials Near Room Temperature Using Portable Emissometers	Table C402.3
C 1549—09	Standard Test Method for Determination of Solar Reflectance Near Ambient Temperature Using A Portable Solar Reflectometer	Table C402.3
D 1003—11e1	Standard Test Method for Haze and Luminous Transmittance of Transparent Plastics	C402.4.2.2
E 283—04	Test Method for Determining the Rate of Air Leakage Through Exterior Windows, Curtain Walls and Doors Under Specified Pressure Differences Across the Specimen	C402.5.1.2.2, Table C402.5.2, C402.5.7
E 408—71(2008)	Test Methods for Total Normal Emittance of Surfaces Using Inspection-meter Techniques	Table C402.3
E 779—10	Standard Test Method for Determining Air Leakage Rate by Fan Pressurization	C402.5
E 903—96	Standard Test Method Solar Absorptance, Reflectance and Transmittance of Materials Using Integrating Spheres (Withdrawn 2005)	Table C402.3
E 1677—11	Standard Specification for an Air-retarder (AR) Material or System for Low-rise Framed Building Walls	C402.5.1.2.2
E 1918—06	Standard Test Method for Measuring Solar Reflectance of Horizontal or Low-sloped Surfaces in the Field	Table C402.3
E 1980—11	Standard Practice for Calculating Solar Reflectance Index of Horizontal and Low-sloped Opaque Surfaces	Table C402.3, C402.3.2
E 2178—13	Standard Test Method for Air Permanence of Building Materials	C402.5.1.2.1
E 2357—11	Standard Test Method for Determining Air Leakage of Air Barriers Assemblies	C402.5.1.2.2

REFERENCED STANDARDS

CRRC
Cool Roof Rating Council
449 15th Street, Suite 200
Oakland, CA 94612

Standard reference number	Title	Referenced in code section number
ANSI/CRRC-1—2012	CRRC-1 Standard	Table C402.3, C402.3.2, C402.3.2.1, Table C407.5.1(1)

CSA
CSA Group
8501 East Pleasant Valley
Cleveland, OH 44131-5516

Standard reference number	Title	Referenced in code section number
AAMA/WDMA/CSA 101/I.S.2/A440—11	North American Fenestration Standard/Specification for Windows, Doors and Unit Skylights	Table C402.5.2
CSA B55.1—2012	Test Method for Measuring Efficiency and Pressure Loss of Drain Water Heat Recovery Units	C404.8
CSA B55.2—2012	Drain Water Heat Recover Units	C404.8

CTI
Cooling Technology Institute
P. O. Box 73383
Houston, TX 77273-3383

Standard reference number	Title	Referenced in code section number
ATC 105 (00)	Acceptance Test Code for Water Cooling Tower	Table C403.2.3(8)
ATC 105S—11	Acceptance Test Code for Closed Circuit Cooling Towers	Table C403.2.3(8)
ATC 106—11	Acceptance Test For Mechanical Draft Evaporative Vapor Condensers	Table C403.2.3(8)
STD 201—11	Standard for Certification of Water Cooling Towers Thermal Performances	Table C403.2.3(8)

DASMA
Door and Access Systems Manufacturers Association
1300 Sumner Avenue
Cleveland, OH 44115-2851

Standard reference number	Title	Referenced in code section number
105—92 (R2004)—13	Test Method for Thermal Transmittance and Air Infiltration of Garage Doors	C303.1.3, Table C402.5.2

DOE
U.S. Department of Energy
c/o Superintendent of Documents
1000 Independence Avenue SW
Washington, DC 20585

Standard reference number	Title	Referenced in code section number
10 CFR, Part 430—1998	Energy Conservation Program for Consumer Products: Test Procedures and Certification and Enforcement Requirement for Plumbing Products; and Certification and Enforcement Requirements for Residential Appliances; Final Rule	Table C403.2.3(4), Table C403.2.3(5), Table C404.2
10 CFR, Part 430, Subpart B, Appendix N—1998	Uniform Test Method for Measuring the Energy Consumption of Furnaces and Boilers	C202

REFERENCED STANDARDS

DOE—continued

Standard reference number	Title	Referenced in code section number
10 CFR, Part 431—2004	Energy Efficiency Program for Certain Commercial and Industrial Equipment: Test Procedures and Efficiency Standards; Final Rules	Table C403.2.3(5), C405.7, C405.8, Table C405.8
10 CFR 431 Subpart B App B	Uniform Test Method for Measuring Nominal Full Load Efficiency of Electric Motors	C403.4.4.4, Table C405.8(1), Table C405.8(2), Table C405.8(3), C405.8(4)
NAECA 87—(88)	National Appliance Energy Conservation Act 1987 [(Public Law 100-12 (with Amendments of 1988-P.L. 100-357)]	Tables C403.2.3(1), C403.2.3(2), C403.2.3(4)

ICC

International Code Council, Inc.
500 New Jersey Avenue, NW
6th Floor
Washington, DC 20001

Standard reference number	Title	Referenced in code section number
IBC—15	International Building Code®	C201.3, C303.2, C402.5.3, C501.4
IFC—15	International Fire Code®	C201.3, C501.4
IFGC—15	International Fuel Gas Code®	C201.3, C501.4
IMC—15	International Mechanical Code®	C403.2.4.3, C403.2.6, C403.2.6.1, C403.2.6.2, C403.2.7, C403.2.8, C403.2.8.1, C403.2.8.2, C403.2.9, C403.4.4, C403.4.4.6, C406.6, C501.4
IPC—15	International Plumbing Code®	C201.3, C501.4
IMPC—15	International Property Maintenance Code®	C501.4
IPSDC—15	International Private Sewage Disposal Code®	C501.4

IEEE

The Institute of Electrical and Electronic Engineers Inc.
3 Park Avenue
New York, NY 10016

Standard reference number	Title	Referenced in code section number
IEEE 515.1—2012	IEE Standard for the Testing, Design, Installation, and Maintenance of Electrical Resistance Trace Heating for Commercial Applications	C404.6.2

IES

Illuminating Engineering Society
120 Wall Street, 17th Floor
New York, NY 10005-4001

Standard reference number	Title	Referenced in code section number
ANSI/ASHRAE/IESNA 90.1—2013	Energy Standard for Buildings, Except Low-rise Residential Buildings	C401.2, Table C402.1.3, Table C402.1.4, C406.2, C502.1, C503.1, C504.1

REFERENCED STANDARDS

ISO
International Organization for Standardization
1, rue de Varembe, Case postale 56, CH-1211
Geneva, Switzerland

Standard reference number	Title	Referenced in code section number
ISO/AHRI/ASHRAE 13256-1 (2011)	Water-to-Air and Brine-to-air Heat Pumps -Testing and Rating for Performance	Table C403.2.3(2)
ISO/AHRI/ASHRAE 13256-2(2011)	Water-to-Water and Brine-to-Water Heat Pumps -Testing and Rating for Performance	C403.2.3(2)

NEMA
National Electrical Manufacturers Association
1300 North 17th Street, Suite 1752
Rosslyn, VA 22209

Standard reference number	Title	Referenced in code section number
MG1—1993	Motors and Generators	C202

NFPA
National Fire Protection Association
1 Batterymarch Park
Quincy, MA 02169-7471

Standard reference number	Title	Referenced in code section number
70—14	National Electrical Code	C501.4

NFRC
National Fenestration Rating Council, Inc.
6305 Ivy Lane, Suite 140
Greenbelt, MD 20770

Standard reference number	Title	Referenced in code section number
100—2009	Procedure for Determining Fenestration Products U-factors—Second Edition	C303.1.3, C402.2.2
200—2009	Procedure for Determining Fenestration Product Solar Heat Gain Coefficients and Visible Transmittance at Normal Incidence—Second Edition	C303.1.3, C402.4.1.1
400—2009	Procedure for Determining Fenestration Product Air Leakage—Second Edition	Table C402.5.2

SMACNA
Sheet Metal and Air Conditioning Contractors National Association, Inc.
4021 Lafayette Center Drive
Chantilly, VA 20151-1209

Standard reference number	Title	Referenced in code section number
SMACNA—2012	HVAC Air Duct Leakage Test Manual 2nd Edition	C403.2.8.1.3

REFERENCED STANDARDS

UL

UL LLC
333 Pfingsten Road
Northbrook, IL 60062-2096

Standard reference number	Title	Referenced in code section number
710—12	Exhaust Hoods for Commercial Cooking Equipment	C403.2.8
727—06	Oil-fired Central Furnaces—with Revisions through April 2010	Table C403.2.3(4)
731—95	Oil-fired Unit Heaters—with Revisions through August 2012	Table C403.2.3(4)
1784—01	Air Leakage Tests of Door Assemblies— with Revisions through July 2009	C402.5.3

US-FTC

United States-Federal Trade Commission
600 Pennsylvania Avenue NW
Washington, DC 20580

Standard reference number	Title	Referenced in code section number
CFR Title 16 (May 31, 2005)	*R*-value Rule	C303.1.4

WDMA

Window and Door Manufacturers Association
2025 M Street, NW, Suite 800
Washington, DC 20036-3309

Standard reference number	Title	Referenced in code section number
AAMA/WDMA/CSA 101/I.S.2/A440—11	North American Fenestration Standard/Specification for Windows, Doors and Unit Skylights	Table C402.5.2

INDEX

A

ACCESSIBLE
    Controls C402.2.4, C404.6, C404.9.1,
                      C405.2.2.3, C405.2.3.1
    Defined C202
ADDITIONAL EFFICIENCY PACKAGE C406
ADDITIONS
    Defined C202
    Historic buildings C501.6
    Requirements C502
ADMINISTRATION . . . . Chapter 1
AIR BARRIER C402.5
    Access openings C402.5.4
    Assemblies C402.5.1.2.2
    Compliance options C402.5.1.2
    Construction C402.5.1.1
    Dampers C402.5.5, C403.2.4.3
    Doors other than fenestration C402.5.4
    Fenestration . . . . C402.5.2, Table C402.5.2
    Materials C402.5.1.2.1
    Penetrations C402.5.1.1
    Recessed lighting C402.5.8
    Rooms with fuel burning appliances C402.5.3
    Testing C402.5
    Vestibules C402.5.7
AIR CONDITIONERS
    Efficiency requirements . . . . . . Tables C403.2.3(1, 3)
AIR CURTAIN
    Defined C202
    Vestibules C402.5.7
AIR ECONOMIZERS
    Defined C202
    Requirements C403.3, C403.3.1, C403.3.2,
                         C403.3.3, C403.3.4
AIR INFILTRATION (see AIR BARRIER)
    Defined C202
AIR INTAKES AND EXHAUSTS C402.5.5,
                                 C403.2.4.3
AIR LEAKAGE—THERMAL ENVELOPE
    (see AIR BARRIER)
AIR SYSTEM BALANCING C408.2.2.1,
                        C408.2.2.2, C408.2.5.3
ALTERATIONS
    Defined C202
    Historic buildings C501.6
    Replacement fenestration C401.2.1
    Requirements C503

ALTERNATE MATERIALS C102
APPROVED
    Defined C202
APPROVED AGENCY
    Defined C202
    Inspections C104.4
AREA-WEIGHTED U-FACTOR C402.4.3.4
AUTOMATIC
    Defined C202

B

BASEMENT WALLS (see WALL, BELOW GRADE)
    Requirements C303.2.1
BELOW-GRADE WALLS
    (see WALL, BELOW GRADE)
BOARD OF APPEALS C109
    Limitations on authority C109.2
    Qualifications of members C109.3
BOILERS
    Defined C202
    Requirements . . . . Table C403.2.3(5), C403.4.2,
                   C403.4.2.5, C404.2, Table C404.2
    Setback controls C403.2.5
    Turndown controls C403.4.2.5
BUILDING
    Defined C202
    Multifamily residential C407.5.2.3
BUILDING COMMISSIONING
    Defined C202
    Requirements C408
BUILDING ENTRANCE
    Defined C202
    Exterior lighting . . . . Table C405.5.2(2)
    Lighting controls C405.2.1.1
    Vestibules C402.5.7
BUILDING ENVELOPE
    Compliance documentation C103.2, C103.2.1
    Defined C202
    Exemptions C402.1.1, C402.1.2
    Insulation C303.1.1
    Insulation and
        fenestration criteria . . . . C402.1.3, Table C402.1.3,
                     C402.1.4, Table C402.1.4
    Requirements C402
    Performance method C407.3
BUILDING THERMAL ENVELOPE
    Air leakage and barriers C402.5

Defined C202
Doors C402.4.4
Low-energy buildings C402.1.1
Performance C402.1, C402.1.3,
C402.1.4, C402.1.5
Rooms with fuel-burning appliances C402.5.3
Specific insulation C402.2

C

C-FACTOR
Defined C202
Assembly U-, C- or F-factor method C402.1.4,
Table C402.1.4

CAULKING AND
WEATHERSTRIPPING ... C402.5.1.1, C402.5.1.2.1,
C402.5.1.2.2, C402.5.3,
C402.5.4, C402.5.6, C402.5.8

CHANGE OF OCCUPANCY C501.4, C505

CHILLERS .... Table C403.2.3(7), Table C407.5.1(4)
Positive displacement
chilling packages C403.2.3.2
Water-cooled centrifugal
chiller packages C403.2.3.1,
Table C403.2.3(7), Table C407.5.1(5)

CIRCULATING HOT WATER SYSTEM
Defined C202
Requirements C404.6.1, C404.6.3

CIRCULATING PUMPS C403.4.2.3

CIRCULATING SYSTEMS C403.4.2.3,
C404.6

CLIMATE TYPES
Defined .... Table C301.3(1)

CLIMATE ZONES
Defined C202
Established .... C301, Figure C301.1,
Table C301.1
By state or territory ... Figure C301.1, Table C301.1
International climate zones C301.3,
Table C301.3(1), Table C301.3(2)
Warm humid ..... C301.2, Table C301.2, C301.3.1
Tropical C301.4

CODE OFFICIAL
Approval of alternate methods C102
Defined C202
Examination of construction documents C103.3
Inspections C104

COEFFICIENT OF PERFORMANCE (COP)
Defined C202
Requirements .... Table C403.2.3(2),
Table C403.2.3(7)

COMMERCIAL BUILDINGS
Compliance C101.2, C101.4.1,
C101.5, C401.1, C401.2, Chapter 5
Defined C202
Total building performance C407

COMMISSIONING C408

COMPLIANCE AND ENFORCEMENT
General C101.5
Existing buildings C501.4

COMPRESSOR (REFRIGERATION)
SYSTEMS C403.5.2

COMPONENT PERFORMANCE
APPROACH C402.1.5

COMPUTER ROOM
Air conditioning .... Table C403.2.3(9)
Defined C202

CONDENSING UNITS
Defined C202
Efficiency requirements ...... Tables C403.2.3(1, 6)

CONDITIONED FLOOR AREA
Defined C202
Renewable energy C406.5

CONDITIONED SPACE
Defined C202
Change from nonconditioned or
low energy C503.2
Roof solar reflectance C402.3
Rooms containing fuel-burning
appliances C402.5.3

CONSTRUCTION DOCUMENTS C103
Amended C103.4
Approval C103.3.1
Examination C103.3
Information required C103.2
Phased approvals C103.3.3
Previous approvals C103.3.2
Retention C103.5
Revocation C104.7.1

CONTINUOUS AIR BARRIER
Defined C202
Required C402.5.1

CONTINUOUS INSULATION
Defined C202
Requirements .... C402.1.3, Table C402.1.3,
C402.2.1, C402.2.3, C402.2.4

CONTROLS
Capabilities C403.2.2, C403.2.4.1, C403.2.4.1.2,
C403.2.4.2.1, C403.2.4.2.2, C403.2.6.1,
C403.4.2.3.1, C403.4.2.6, C403.4.3, C404.6, 404.7
Chilled water plants C403.4.2.6
Economizers C403.3, C403.3.1, C403.3.3.2,
C403.3.3.3, Table C403.3.3.3

INDEX

Energy recovery systems C403.2.7
Fan speed. C403.4.1, C403.4.1.1, C403.4.3,
C403.2.12, C403.2.12.1
Freeze protection system C403.2.4.6
Glazing C402.4.3.3
Heat pump C403.2.4.1.1, C403.4.2.3
Heating and cooling C403.2.2, C403.2.4,
C403.3.1, C403.4
Hot water system C404.6
Humidity C403.2.4.1, C403.2.7, C403.3.1
HVAC C403.2.4, C403.2.12, C408.2.3.2
Hydronic systems C403.4.2
Lighting C402.4, C402.4.1.1, C402.4.1.2,
C402.4.2.1, C402.4.3.1, C405.2, C405.5.1
Lighting, digital C406.4
Off hour. C403.2.4.2
Service water heating C403.4.6,
C404.5, C404.6
Shutoff dampers . . . . C403.2.4.3, C 403.4.4
Snow melt system C403.2.4.5
Temperature C403.2.4.1, C403.2.4.2,
C403.2.4.2.1, C403.2.4.2.2,
C403.2.4.2.3, C403.2.4.3 C403.4.2
Three-pipe system C403.4.2.1
Two-pipe changeover system. C403.4.2.2
Variable air volume systems. . . . C403.4.4, C403.5.5
Ventilation. C403.2.6
COOLING SYSTEMS
Hot gas bypass limitation C403.4.6
COOLING TOWER C403.4.3.3, C403.4.3.4
COOLING WITH OUTDOOR AIR C403.3.1
CRAWL SPACE WALLS
Defined C202
Requirements C303.2.1
CURTAIN WALL
Defined C202
Air leakage of fenestration . . . . Table C402.5.2

D

DAMPERS. C402.5.5, C403.2.4.3
DAYLIGHT RESPONSIVE CONTROL
Defined. C202
Required. C402.4.1.1, C402.4.1.2, C402.4.2.1,
C402.4.3.1, C402.4.3.2, C405.2.3.1
DAYLIGHT ZONE C405.2.3.2, C405.2.3.3
Defined C202
Under skylights . . . C402.4.1.2, C402.4.2, C405.2.2.3
DAYLIGHT ZONE CONTROL. C405.2.3
DEADBAND C403.2.4.1.2, C403.2.4.1.3,
C403.4.2.3.1
DEFINITIONS . . . . Chapter 2

DEGREE DAY COOLING (CDD) . . . . Table C301.3(2)
DEGREE DAY HEATING (HDD). . . . . Table C301.3(2)
DEMAND CONTROL VENTILATION (DCV)
Defined C202
Requirements C403.2.5.1
DEMAND RECIRCULATION WATER SYSTEM
Defined C202
Requirements C404.7
DESIGN CONDITIONS C302
DIRECT EXPANSION (DX) C403.4.1.1
DOORS
Default *U*-factors . . . . Table C303.1.3(2)
Garage doors C303.1.3
Loading docks. C402.5.6
Opaque. C402.2.7
Performance requirements . . . . Table C402.1.3,
Table C402.1.4, C402.4,
C402.4.4, 402.5.4
Vestibules C402.5.7
DRAIN WATER HEAT RECOVERY C404.8
DUAL DUCT VAV C403.4.4.2, C403.4.4.3
DUCTS
Defined C202
Insulation. C103.2, C403.2.9, C403.2.9.1,
C403.2.9.1.1, C403.2.9.1.2, C403.2.9.1.3
Sealing C103.2, C403.2.9, C403.2.9.1.1,
C403.2.9.1.2, C403.2.9.1.3
DUCT SYSTEM
Defined C202
Requirements C403.2.9.1
DWELLING UNIT
Defined C202
Electrical Meter. C405.6
Lighting C405.1
Vestibules C402.5.7
DYNAMIC GLAZING
Defined C202
Requirements C402.4.3.3

E

ECONOMIZER
Air . . . . C403.3.3, Table C403.3(1)
Controls. C403.3.1, C403.4.1.2, C403.4.1.3
Defined C202
Fault detection and diagnostics (FDD) . . . C403.2.4.7
High-limit shutoff control C403.3.3.3,
Table C403.3.3.3
Requirements C403.3.3, C403.3.4
Water C403.3.4
EFFICIENCY, ADDITIONAL C406
ELECTRICAL METERS C405.6

ELECTRICAL MOTORS..................C405.8
ELECTRICAL POWER AND LIGHTING.......C405
ELECTRICAL TRANSFORMERS..........C405.7
ELEVATOR POWER..........C405.9.1, C405.9.2
ELEVATOR SHAFTS..........C402.5.4, C402.5.5
ENCLOSED SPACE
   Defined.................................C202
   Under skylights......................C402.4.2
ENERGY ANALYSIS, ANNUAL
   Defined.................................C202
   Documentation........................C407
   Requirements..........................C407
ENERGY COST
   Compliance performance..................C401.2
   Defined.................................C202
   Performance basis......................C407.3
**ENERGY EFFICIENCY RATIO
(EER)**......C403.2.3, Tables C403.2.3(1, 2, 3, 6, 7)
ENERGY RECOVERY VENTILATION SYSTEMS
   Defined.................................C202
   Requirements...............C403.2.1, C403.2.7
ENERGY SIMULATION TOOL
   Defined.................................C202
   Requirements/use............... C101.5.1, C407,
                         C407.2, C407.5
ENTRANCE DOOR
   Air leakage..................... Table C402.5.2
   Defined.................................C202
   Thermal performance..............Table C402.4
ENVELOPE, BUILDING THERMAL
   Defined.................................C202
ENVELOPE DESIGN METHODS.........C402.1.3,
                        C402.1.4, C402.1.5
EQUIPMENT BUILDINGS.................C402.1.2
EQUIPMENT EFFICIENCIES......C103.2, C403.2.3,
                 C403.3.1, C403.4.1, C404.2
**EQUIPMENT PERFORMANCE
REQUIREMENTS**.....................C403.2.3
   Boilers..................... Table C403.2.3(5)
   Condensing units............ Table C403.2.3(6)
   Economizer exception....... Tables C403.3.1(1-6)
   Heat rejection equipment....... Table C403.2.3(8)
   Packaged terminal air conditioners and
     heat pump.................. Table C403.2.3(3)
   Unitary air conditioners and
     condensing units............. Table C403.2.3(1)
   Unitary and applied heat pumps ..Table C403.2.3(2)
   Warm air duct furnaces and
     unit heaters................Table C403.2.3(4)
   Warm air furnaces............Table C403.2.3(4)
   Warm air furnaces/air-conditioning
     units......................Table C403.2.3(4)
   Water chilling packages,
     standard..................Table C403.2.3(7)
   Water heating............. Table C404.2, C404.2.1
EQUIPMENT ROOM
   Defined.................................C202
   Lower power allowance.........Table C405.5.2(2)
ESCALATORS.........................C405.9.2
EXEMPT BUILDINGS..........C401.1.1, C402.1.2
EXHAUSTS.............................C402.5.5
EXIT SIGNS............. C405.2, C405.3, C405.4.1
EXISTING BUILDINGS.................Chapter 5
EXTERIOR LIGHTING............C405.2.5, C405.5
EXTERNAL SHADING............Table C407.5.1(1)
EXTERIOR WALLS
   Defined.................................C202
   Thermal performance..............C402, C402.2.3

F

F–FACTOR
   Defined.................................C202
   Assembly U-, C- or F-factor
     method................C402.1.4, Table C402.1.4
FAN BRAKE HORSEPOWER (BHP)
   Allowable...........................C403.2.12
   Defined.................................C202
FAN EFFICIENCY GRADE (FEG)
   Defined.................................C202
   Requirements.....................C403.2.12.3
FAN FLOOR HORSEPOWER.........C403.2.12.1
**FAN POWER
LIMITATION**............. Tables C403.2.12.1(1, 2)
FAN SYSTEM BHP
   Allowable..........................C403.2.12.1
   Defined.................................C202
FAN SYSTEM DESIGN CONDITIONS
   Allowable..........................C403.2.12.2
   Defined.................................C202
FAN SYSTEM MOTOR NAMEPLATE HP
   Defined.................................C202
   Requirements......................C403.2.12
FAULT DETECTION & DIAGNOSTICS (FDD)
   Economizers.....................C403.2.4.7
FEES...................................C107
   Refunds............................C107.5
   Related fees........................C107.4
   Schedule of permit fees.................C107.2
**FENESTRATION
(see also Doors)**.............C303.1.3, C402.4
   Air leakage (infiltration) rate............C402.5.2,
                            Table C402.5.2
   Defined.................................C202

INDEX

Maximum area C402.4.1, C402.4.1.2
Rating and labeling. C303.1.3, C402.1.3
Skylights C402.4.1.2, C402.4.2,
C402.4.2.1, C402.4.2.2, C402.4.3,
C502.2.2, C503.3.3
Solar heat gain (SHGC) . . . . C402.4.3, Table C402.4
Vertical C402.1, C402.4.1.1, C402.4.3,
Table C402.4, C502.2.1, C503.3.2

FENESTRATION PRODUCT,
   FIELD-FABRICATED
   Defined C202
   Air leakage C402.5.2

FENESTRATION PRODUCT, SITE-BUILT
   Defined C202

FLOOR AREA, NET
   Defined C202
   Fenestration increase C402.4.1.1

FLOORS
   Slab on grade C402.2.5
   Thermal properties .C402.2.4

FREEZE PROTECTION SYSTEMS C403.2.4.6

FURNACE EFFICIENCY. . . . . Table C403.2.3(4)

G

GENERAL LIGHTING
   Additional lighting C405.4.2.1.1
   Daylight controls C405.2.3
   Defined C202
   Interior lighting power C405.4.1

GENERAL PURPOSE ELECTRIC MOTORS
   Defined C202

GREENHOUSE
   Defined C202
   Building envelope C402.1.1

GUESTROOMS (see SLEEPING UNIT)

GLAZING AREA
   Default fenestration *U*-factors . . . . Table C303.1.3(1)
   Dynamic C402.4.3.3

H

HAZE FACTOR C402.4.2.2
HEAT PUMP . . . . .Tables C403.2.3(2, 3),
C403.2.4.1.1, C403.4.2.3

HEAT RECOVERY
   Drain water C404.8, C406.7.1
   Economizer exemption C403.3
   Kitchen exhaust C403.2.8
   Service water C403.4.5

HEAT REJECTION EQUIPMENT C403.4.3,
Table C403.2.3(8)
HEAT TRACE SYSTEMS C404.6.2
HEAT TRANSFER
   EQUIPMENT. . . . . Table C403.2.3(9)
HEAT TRAPS
   Defined C202
   Required C404.3, C404.4
HEATED SLAB
   Defined C202
   Insulation. . . . . Table C402.1.3,
Table C402.1.4, C402.2.6
HEATING AND COOLING
   LOADS C302.1, C403.2.1, C403.2.2,
C403.2.3, C403.2.4.1.1, C403.3
HEATING OUTSIDE A BUILDING C403.2.13
HIGH-SPEED DOOR
   Air leakage . . . . Table C402.5.2
   Defined C202
HISTORIC BUILDINGS
   Defined C202
   Compliance. C501.6
HOT GAS BYPASS. . . . . . . C403.4.6, Table C403.4.6
HOT WATER C404.2, C404.6
   Efficient delivery C404.5
   Piping insulation C403.2.10, C404.4
   System controls C403.4.5, C404.6
HUMIDISTAT
   Defined C202
   Requirements C403.2.4.1, C403.2.7
HVAC EQUIPMENT
   Automatic setback and shutdown C403.2.4.2.2
   Automatic start capabilities C403.2.4.2.3
   Increased efficiency C406.2
   Performance requirements C403.2.3
   Supply-air temperature reset C403.4.4.5
   System map zones . . . . Table C407.5.1(2),
C407.5.2.1
HVAC SYSTEMS. C403.2.11, C408.2
   Manuals C408.2.5.2
   Plan. C408.2.1
   Report C408.2.4, C408.2.4.1,
C408.2.4.2, C408.2.5.4
HYDRONIC HEAT PUMP SYSTEMS C403.4.2.3

I

ICE MELT SYSTEMS C403.2.4.5
IDENTIFICATION (MATERIALS,
   EQUIPMENT AND SYSTEM). C303.1

INDEX

INDIRECTLY CONDITIONED SPACE
(see CONDITIONED SPACE)

INFILTRATION (air leakage)
(see AIR BARRIER)
    Defined C202

INSPECTIONS C104
    Inspection agencies C104.4

INSULATED SIDING C303.1.4.1

INSULATION
    Continuous insulation ... C402.1.3, Table C402.1.3, C402.2.1, C402.2.2
    Duct C403.2.9
    Identification C303.1, C303.1.2
    Installation C303.1.1, C303.1.1.1, C303.1.2, C303.2
    Mechanical system piping C403.2.10
    Piping C404.4
    Plenum C403.2.9
    Product rating C303.1.4
    Protection of exposed foundation C303.2.1
    Protection of piping insulation C403.2.10.1
    Radiant heating systems C402.2.6
    Requirements .... C402.1.3, Table C402.1.3, C402.2 through C402.2.6

INTEGRATED PART LOAD VALUE (IPLV)
    Defined C202
    Requirements .... Tables C403.2.3(6, 7), C403.2.3.1, Table 403.3(2)

INTERIOR LIGHTING POWER C405.4, C405.4.2

K

KITCHEN EXHAUST ...... C403.2.8, Table C403.2.8

L

LABELED
    Dampers C403.2.4.3
    Defined C202
    Fans C403.2.12.3
    Glazing, skylights C402.2.2, C402.4.2.4, C402.5.2
    HVAC equipment .... Table C403.2.3(3)
    Lighting C402.5.8, C405.4.1
    Requirements .... C303.1.3, C303.3, Table C403.2.3(3)

LIGHTING POWER
    Additional lighting C405.4.2.2.1, C405.5.1
    Design procedures C405.4.2, C405.4.2.1, C405.4.2.2
    Exterior connected .... C405.5, C405.5.1, Tables C405.5.1(1, 2)
    Interior connected .... C405.4, C405.4.1, C405.4.2, Tables C405.4.2(1, 2), C406.3
    Reduced lighting power density C406.3

LIGHTING SYSTEMS C405, C406.3
    Controls, exterior .... C405.2.5, Tables C405.5.1(1, 2), C405.5.1
    Controls, interior ... C405.2.2, Tables C405.4.2(1, 2), C405.4, C405.4.1, C405.4.2, C406.4
    Daylight responsive C405.2.3, C405.2.3.1
    Dwelling and sleeping units C405.1, C405.2.2, C405.2.3, C405.2.4, C405.4.1
    Existing buildings C502.1, C502.2.6, C503.1, C503.6, 504.1
    Light reduction C405.2.2.2
    Occupant sensor controls C405.2.1, C405.2.1.1, C405,2.1.2
    Recessed C402.5.8
    Retail display C405.4.2.2.1
    Specific applications C405.2.4
    Time switch controls C405.2.2, C405.2.2.1

LINER SYSTEM (Ls)
    Defined C202
    Insulation .... Table C402.1.3

LISTED
    Defined C202
    Kitchen exhaust hoods C403.2.8
    Skylights C402.2.2

LOADING DOCK WEATHERSEALS C402.5.6

LOW-ENERGY BUILDINGS C402.1.1

LOW-SLOPED ROOF
    Defined C202
    Roof solar reflectance C402.3

LUMINAIRE
    Controls C405.2, C405.2.1, C405.2.2, C405.2.3, C405.2.4
    Sealed C402.5.8
    Wattage C405.4.1, C405.5.1

M

MAINTENANCE
    General C501.2, C501.3
    Instructions for equipment and systems C303.3
    Owner responsibility C501.3

MANUALS C101.5.1
    Defined C202

MASS
    Floor .... Table C402.1.3, Table C402.1.4
    Wall C402.2.3

MATERIALS AND EQUIPMENT C303

MECHANICAL SYSTEMS AND
  EQUIPMENT C403
  Existing buildings C501.5, C502.1,
        C502.2.3, C503.4, C504
MECHANICAL VENTILATION C403.1, C403.2.6
METERS, ELECTRICAL C405.6
MOTOR NAMEPLATE
  HORSEPOWER C403.2.12.2
  Efficiency C405.8
MOVING WALKWAYS C405.9.2, C405.9.3
MULTIFAMILY
  RESIDENTIAL BUILDINGS C407.5.2.3
MULTIPLE ZONE SYSTEMS C403.4, C403.4.4,
        C403.4.4.5

N

NAMEPLATE HORSEPOWER
  Allowable C403.2.12
  Defined C202
NET FLOOR AREA (see FLOOR AREA, NET)
NONCONDITIONED SPACE
  Alterations C503.2
NONSTANDARD PART LOAD VALUE
  Defined C202

O

OCCUPANCY
  Complex HVAC systems C403.4.4
  Compliance C101.4, C101.5
  Lighting power
    allowances C405.4.2, C405.4.2.1,
        C405.2.2, Tables C405.4.2(1, 2)
  Mixed occupancies C101.4.1
OCCUPANT SENSOR CONTROL
  Commissioning C408.3.1.1
  Defined C202
  Lighting C406.4
  Outdoor heating C403.2.13
  Required C405.2.1, C405.2.1.1, C405.2.1.2
OFF-HOUR, CONTROLS C403.2.4.3, C405.2.2.1
ON-SITE RENEWABLE ENERGY
  Defined C202
  Efficiency package C406.5
OPAQUE AREAS . . . . C402.1, Table C402.1.3,
        Table C402.1.4
OPAQUE DOORS
  Defined C202
  Regulated . . . . . Table C402.1.3, C402.1.4, C402.4.4

OPERATING &
  MAINTENANCE MANUAL C408.2.5.2
ORIENTATION
  Daylight Responsive controls C405.2.3.1
  Fenestration . . . . Table C402.4
  Thermostatic controls C403.2.4.1, C407.5.2.2

P

PACKAGED TERMINAL
  AIR CONDITIONER (PTAC)
  Requirements . . . . Table C403.2.3(3)
PACKAGED TERMINAL HEAT PUMP
  Requirements . . . . Table C403.2.3(3)
PARKING GARAGE VENTILATION C403.2.6.2
PERFORMANCE ANALYSIS C407
PERMIT (see FEES)
  Work commencing before permit C107.3
PIPE INSULATION C403.2.10,
        Table C403.2.10, C404.4
PLANS AND SPECIFICATIONS C103
PLENUMS
  Insulation and sealing C403.2.9
POOLS C404.9
  Controls C404.9.1, 404.9.2
  Covers C404.9.3
  Existing buildings C502.2.5
POWER DENSITY C406.3
POWERED ROOF/WALL VENTILATORS
  Defined C202
  Fan efficiency exemption C403.2.12.3
PROPOSED DESIGN
  Defined C202
  Requirements C407
PUBLIC LAVATORY C404.5
PUMPING SYSTEMS C403.4.2.3, C404.6,
        C408.2.2.2

R

R-VALUE
  Above-grade walls C402.2.3
  Defined C202
  Insulation component method C402.1.3,
        Table C402.1.3
  Roof assemblies C402.2.2
  Slabs on grade C402.2.5
  Steel stud walls . . . . . . C402.1.4.1, Table C402.1.4.1
RADIANT HEATING SYSTEM
  Defined C202
  Insulation C403.3.6

INDEX

READILY ACCESSIBLE
  Defined C202
  Lighting controls.......... C405.2.2.3, C405.2.3.1
RECOOLING C403.4.4
REFERENCED STANDARDS ...... C106, Chapter 6
REFRIGERATED WAREHOUSE COOLER
  Defined C202
  Requirements C403.2.15
REFRIGERATED WAREHOUSE FREEZER
  Defined C202
  Requirements C403.2.15
REFRIGERATION EQUIPMENT
  Performance .... C403.2.14, Tables C403.2.14(1, 2)
REGISTERED DESIGN PROFESSIONAL
  Commissioning C408
  Defined C202
REHEATING........ C403.4.5, C403.4.4.5, C403.4.5
RENEWABLE/NONDEPLETABLE
ENERGY SOURCES.................... C406.5
REPAIR
  Defined C202
  Historic buildings C501.6
  Requirements C501.5, C504
REPLACEMENT MATERIALS............. C501.5
  Replacement fenestration C401.2.1
REROOFING
  Defined C202
RESET CONTROL................... C403.4.4.5
RESIDENTIAL BUILDINGS
  Compliance C101.2, C101.4.1, C101.5
  Defined C202
ROOF ASSEMBLY
  Air barriers....................... C402.5.1.2.1
  Defined C202
  Fenestration...... C402.4.1, C402.4.1.2, C402.4.2,
                                          C405.2.3.3
  Recover C503.1
  Reflectance and
    emittance options .... Table C402.3
  Repairs C504.1
  Replacement C503.3.1
  Requirements C303.1.1.1, C402.2.2
  Solar reflectance and
    thermal emittance.................... C402.3
ROOF RECOVER
  Defined C202
  Exemption C503.1
ROOF REPAIR
  Defined C202
  Exemption C504.1

ROOF REPLACEMENT
  Defined................................. C202
  Requirements....................... C503.3.1
ROOFTOP MONITOR
  Daylight zones C405.2.3.2
  Defined................................. C202
  Skylights required.................... C402.4.2
ROOF VENTILATORS
  (see POWERED ROOF/ WALL VENTILATORS)
ROOMS WITH FUEL-BURNING
APPLIANCES C402.5.3

S

SATURATED CONDENSING TEMPERATURE
  Defined................................. C202
  Refrigeration systems C403.5.1
SCOPE OF CODE C101.2
SCREW LAMP HOLDERS
  Defined................................. C202
  Requirements....................... C405.4.1
SEASONAL ENERGY EFFICIENCY
RATIO (SEER).......... Tables C403.2.3(1, 2, 3)
SERVICE WATER HEATING
  Commissioning....................... C404.11
  Defined................................. C202
  Drain water heat recovery C404.8
  Efficiency C404.2.1, 404.5
  Existing buildings C502.2.4, C503.5, C504.1
  Reduced use C406.7
  Requirements.......... C403.4.5, C404, C404.2,
                          C404.2.1, 404.5, 404.6, 404.7
SETBACK THERMOSTAT............. C403.2.4.2,
                                  C403.2.4.2.1, C403.2.4.2.2
SHADING C402.3, C402.4.3
SHGC
  (see SOLAR HEAT GAIN COEFFICIENT)
SHUTOFF DAMPERS C402.5.5, C403.2.4.3
SIMULATED PERFORMANCE
ALTERNATIVE C407
SIMULATION TOOL
  (see ENERGY SIMULATION TOOL)
SINGLE ZONE C403.3
SIZING
  Equipment and system................ C403.2.2
SKYLIGHTS C402.1.5, C402.3,
                  Table C402.4, C402.4.3.1, C402.4.3.2
  Additions C502.2.2
  Air leakage (infiltration) .... Table C402.5.2
  Alterations C503.3.3
  Curb insulation...................... C402.2.2
  Defined (see Fenestration)............... C202

Haze factor C402.4.2.2
Lighting controls C402.4.2.1
Maximum area C402.4.1, C402.4.1.2
Minimum area. C402.4.2
SLAB-EDGE INSULATION. C303.2.1, C402.2.5
SLEEPING UNIT
  Defined. C202
  Lighting. C405.1, C405.2.2, C405.2.4, C405.2.3
SMALL ELECTRIC MOTOR
  Defined. C202
  Minimum efficiency. C405.8
**SNOW MELT AND ICE SYSTEM
  CONTROLS** C403.2.4.5
**SOLAR HEAT GAIN COEFFICIENT
  (SHGC)**. . . . C103.2, Table C303.1.3(3), C402.4.1.1,
            C402.4.3, Table C402.4, C402.4.3.1
  Defined. C202
  Dynamic glazing C402.4.3.3
  Replacement products C401.2.1
SPAS. C404.9, C404.10
STAIRWAYS C402.5.4, C402.5.5, C403.2.4.3,
            C405.2, C405.2.1.1, Table C405.4.2(2)
STANDARD REFERENCE DESIGN
  Defined. C202
  Requirements . . . . C407, Tables C407.5.1(1, 3)
STANDARDS, REFERENCED . . . . . . C106, Chapter 6
STEEL FRAMING . . . . Table C402.1.3,
            Table C402.1.4, C402.1.4.1
STOP WORK ORDER. C108
  Authority C108.1
  Emergencies. C108.3
  Failure to comply C108.4
  Issuance C108.2
STOREFRONT
  Defined. C202
  Glazing . . . . .Table C402.5.2
**SUPPLY AIR TEMPERATURE
  RESET CONTROLS** C403.4.5.5
SUSPENDED CEILINGS. C402.2.2
SWIMMING POOLS C404.9

T

TEMPERATURE DEAD BAND. C403.4.3.3.1
TENANT SPACES. C406.1.1
THERMAL CONDUCTANCE (see C-Factor)
THERMAL MASS (see MASS)
THERMAL RESISTANCE (see *R*-VALUE)
THERMAL TRANSMITTANCE (see *U*-FACTOR)
THERMOSTAT
  Defined. C202
  Pools and spa heaters C404.9.1

Requirements C403.2.4, C403.2.4.1,
            C403.2.4.1.2, C403.2.4.1.3,
            C403.2.4.2, C403.4.4
  Setback capabilities C403.2.4.2
TIME SWITCH CONTROL
  Defined. C202
  Requirements C405.2.2, C405.2.2.1
TOTAL BUILDING PERFORMANCE C407
TOWNHOUSE (see RESIDENTIAL BUILDINGS)
TRANSFORMERS, ELECTRIC. C405.7

U

***U*-FACTOR**
  Area-weighted *U*-factor fenestration C402.4.3.4
  Assembly U-, C- or F-factor
    method. . . . . C402.1.4, Table C402.1.4
  Component performance alternative. C402.1.5
  Default door . . . . Table C303.1.3(2)
  Default glazed fenestration . . . . . . Table C303.1.3(1)
  Defined C202
  Fenestration . . . . C401.2.1, C402.4.3, Table C402.4
  Skylights . . . . Table C402.4, C402.4.3.2
  Steel framing. C402.1.4.1

V

**VARIABLE AIR VOLUME SYSTEMS
  (VAV)** C403.3.2, C403.4.1.2, C403.4.5,
            C403.4.4.1, C403.4.4.2, C403.4.5.4,
            C403.4.5.6, Table C407.5.1(3)
VARIABLE REFRIGERANT FLOW SYSTEM
  Defined C202
  Increased efficiencies C406.2
VENTILATION. C403.2.6
  Defined C202
  Demand controlled ventilation (DCV) C403.2.6.1
  Energy recovery system C403.2.7, C406.6
  Parking garages C403.2.6.2
VENTILATION AIR
  Defined C202
  Energy recovery C403.2.7
  Fan controls C403.4.1.1
  Kitchen exhaust C403.2.8
VERTICAL FENESTRATION (see FENESTRATION)
VESTIBULES. C402.5.7
VISIBLE TRANSMITTANCE (VT)
  Default glazed fenestration . . . . . . Table C303.1.3(3)
  Defined C202
  Dynamic glazing C402.4.3.3
  Increased fenestration C402.4.1.1
  Skylights C402.4.2, C405.2.3.3

W

WALK-IN COOLER
  Defined C202
  Requirements C403.2.15, C403.2.16

WALK-IN FREEZER
  Defined C202
  Requirements C403.2.15, C403.2.16

WALL
  Above-grade wall, defined C202
  Thermal resistance ..... C402.2.3, Table C402.1.3,
  Table C402.1.4
  Below-grade wall, defined C202
  Thermal resistance .... Table C402.1.3,
  Table C402.1.4, C402.1.5
  Crawlspace wall, defined C202
  Exterior wall, defined C202
  Steel framed .... C402.1.4.1, Table C402.1.4.1

WALLS (see EXTERIOR WALLS AND ENVELOPE, BUILDING THERMAL)

WALLS ADJACENT TO UNCONDITIONED SPACE (see BUILDING THERMAL ENVELOPE)

WALL VENTILATORS (see POWERED ROOF/ WALL VENTILATORS)

WATER ECONOMIZER C403.3, C403.3.4,
C403.3.4.1, C403.3.4.2

WATER HEATER
  Defined C202
  Efficiency .... Table C404.2, C404.2.1

WATER HEATING .... C404, Table C404.2

WINDOW AREA (see FENESTRATION and GLAZING AREA)

Z

ZONE (see also CLIMATE ZONES)
  Defined C202
  Requirements C403.2.4, C403.3,
  C403.4, C407.5.2.1, C407.5.2.2

EDITORIAL CHANGES – SECOND PRINTING

Page vii: paragraph 3, line 8 now reads . . .visions therefore contain provisions for residential buildings four stories or greater in height. Each

Page C-16: column 5, last line now reads . . .3A Coweta

Page C-31, Section C402.1: Item 1, line 6 now reads . . .component performance alternative of Section C402.1.5.

Page C-32, Section C402.1.3: line 23 now reads . . .shall comply with Table C402.1.4 and opaque nonswinging

Page C-32, Section C402.1.4: last 2 lines now read . . .doors shall comply with Table C402.1.4 and opaque nonswinging doors shall comply with Table C402.1.3.

Page C-33, TABLE C402.1.3: column 1, row 4 now reads . . .Metal buildings[b]

Page C-33, TABLE C402.1.3: column 1, row 18 now reads . . .Heated slabs

Page C-35, Section C402.2.3: line 5 now reads . . .C402.1.3, based on framing type and construction materi-

Page C-35, Section C402.2.3: Item 3, line 2 now reads . . .kJ/m^2 · K).

Page C-36, Section C402.3.1: equation notation $R_{initial}$ now reads . . $R_{initial}$ = The initial solar reflectance determined in accordance with CRRC-1 Standard.

Page C-36, TABLE C402.3: Note a, line 2 now reads . . .reflectance or thermal emittance shall be assigned both a 3-year-aged solar reflectance in accordance with Section C402.3.1 and a 3-year-aged thermal

Page C-36, TABLE C402.3: Note b now reads . . .Aged solar reflectance tested in accordance with ASTM C 1549, ASTM E 903 or ASTM E 1918 or CRRC-1 Standard.

Page C-36, TABLE C402.3: Note c now reads . . .Aged thermal emittance tested in accordance with ASTM C 1371 or ASTM E 408 or CRRC-1 Standard.

Page C-36, TABLE C402.4: column 9 now reads . . .

8	
0.29	
0.37	
0.77	
SEW	N
0.45	N
NR	NR
NR	NR
0.50	
NR	

Page C-38, Section C402.4.3.3: Exception, line 4 now reads . . .requirements of Table C402.4.

Page C-38, Section C402.5.1.1: Item 4, line 2 now reads . . .tion C402.5.8. Where similar objects are installed

Page C-39, Section C402.5.4: Item 2, line 1 now reads . . .2. Doors and door openings required to comply with

Page C-45, TABLE C403.2.3(3): column 1, row 12 now reads . . .Room air conditioners, without louvered sides

Page C-48, TABLE C403.2.3(7): column 2, row 2 now reads . . .< 150 Tons

Page C-48, TABLE C403.2.3(7): column 2, row 10 now reads . . .< 150 Tons

Page C-48, TABLE C403.2.3(7): column 4, row 18 now reads . . .≤ 0.634 FL

Page C-48, TABLE C403.2.3(7): column 4, row 19 now reads . . .≤ 0.596 IPLV

Page C-50, TABLE C403.2.3(9): column 2, row 2 now reads . . .< 65,000 Btu/h

Page C-50, TABLE C403.2.3(9): column 2, row 5 now reads . . .< 65,000 Btu/h

Page C-50, TABLE C403.2.3(9): column 2, row 8 now reads . . .< 65,000 Btu/h

Page C-50, TABLE C403.2.3(9): column 2, row 11 now reads . . .< 65,000 Btu/h

Page C-50, TABLE C403.2.3(9): column 2, row 14 now reads . . .< 65,000 Btu/h

Page C-55, Section C403.2.9.1.3: line 5 now reads . . .C403.2.9. In addition, ducts and plenums shall be

Page C-55, Section 403.2.10.1 has been added.

Page C-55, Section C403.2.12.1 title now reads . . .**C403.2.12.1 Allowable fan motor horsepower.**

Page C-56, TABLE C403.2.1 title now reads . . .**TABLE C403.2.10 MINIMUM PIPE INSULATION THICKNESS (in inches)**[a, c]

Page C-56, TABLE C403.2.1: column 2, row 2 now reads . . .Conductivity Btu · in./(h · ft^2 · °F)[b]

Page C-56, TABLE C403.2.1: column 8, row 2 now reads . . .≥ 8

Page C-61, Section C403.3: line 2 now reads . . .shall include either an air or water economizer complying

Page C-61, Section C403.3: Item 2, line 2 now reads . . .vidual fan cooling units have a capacity of less than

Page C-61, Section C403.3: Item 9, line 2 now reads . . .accordance with Section C403.4.5.

Page C-62, TABLE C403.3.1 title now reads . . .**TABLE C403.3.1 DX COOLING STAGE REQUIREMENTS FOR MODULATING AIRFLOW UNITS**

Page C-64, Section C403.4.1.3: line 2 now reads . . .systems with direct digital control of individual zones

Page C-76, Section 405.4.2.2.1 title now reads . . .**C405.4.2.2.1 Additional interior lighting power.**

Page C-77, Section 405.4.2.2.1: Item 2, line 6 now reads . . .than 1.0 w/ft^2 (10.7 w/m^2) of such spaces.

Page C-77, TABLE C405.4.2(1): column 1, row 3 now reads . . .Convention center

Page C-89, TABLE C407.5.1(3): Note a, line 3 now reads . . .Section C403.4.4, Exception 4. Supply air temperature setpoint shall be constant at the design condition.

Page C-89, TABLE C407.5.1(3): Note e, line 11 now reads . . .C403.4.3.3. The heat rejection device shall be an axial fan cooling tower with two-speed fans where required in Section C403.4.3. Condenser water design

EDITORIAL CHANGES – THIRD PRINTING

Page C-31, Section C401.2: Item 3, line 2 now reads . . .C405.2, C405.3, C405.5, C405.6 and C407. The build-

Page C-31, Section C401.2.1: line 6 now reads . . .Table C402.4.

Page C-31, Section C401.2.1: Exception line 4 now reads . . .listed in Table C402.4 shall be permitted to satisfy the

Page C-31, Section C401.2.1: Exception line 6 now reads . . .category listed in Table C402.4. Individual fenestration

Page C-35, Section C402.2.4: Exception 1, line 7 now reads . . .C402.1.3 for "Metal framed" or "Wood framed

Page C-37, Section C402.4.2: Exception 5, line 5 now reads . . .C405.2.3.

Page C-37, Section C402.4.2.1: line 4 now reads . . .electric lights within *daylight zones* under skylights.

Page C-39, Section C402.5.4: Exception 1, line 2 now reads . . .716 or 716.5 of the *International Building Code*.

Page C-40, Section C402.5.7: Exception 6 has been renumbered as Exception 7 and a new Exception 6 now reads . . .Doors used primarily to facilitate vehicular movement or material handling and adjacent personnel doors.

Page C-53, Section C403.2.6.1: line 5 now reads . . .of floor area (as established in Table 403.3.1.1 of the

Page C-61, Section C403.3: Exception 2 now reads as shown:

2. In *climate zones* other than 1A and 1B, where individual fan cooling units have a capacity of less than 54,000 Btu/h (15.8 kW) and meet one of the following:

    2.1. Have direct expansion cooling coils.

    2.2. The total chilled water system capacity less the capacity of fan units with air economizers is less than the minimum specified in Table C403.3(1).

    The total supply capacity of all fan-cooling units not provided with economizers shall not exceed 20 percent of the total supply capacity of all fan-cooling units in the building or 300,000 Btu/h (88 kW), whichever is greater.

Page C-75, Section C405.2.4: Exception now reads as shown:

3. Hotel and motel sleeping units and guest suites shall have a master control device that is capable of automatically switching off all installed luminaires and switched receptacles within 20 minutes after all occupants leave the room.

    Exception: Lighting and switched receptacles controlled by captive key systems.

Page C-79, Section C405.5.1: line 5 now reads . . .ted in Table C405.5.1(2) for the applicable lighting zone.

Page C-79, Section C405.5.1: line 7 now reads . . .tions listed in Table C405.5.1(2), in the Tradable Surfaces sec-

Page C-79, Section C405.5.1: line 9 now reads . . .from Table C405.5.1(1) unless otherwise specified by the

Page C-79, TABLE C405.5.2(1): title now reads . . .**TABLE C405.5.1(1) EXTERIOR LIGHTING ZONES**

Page C-80, TABLE C405.5.2(2): title now reads . . .**TABLE C405.5.1(2) INDIVIDUAL LIGHTING POWER ALLOWANCES FOR BUILDING EXTERIORS**

IECC—RESIDENTIAL PROVISIONS

TABLE OF CONTENTS

CHAPTER 1 SCOPE AND ADMINISTRATION R-3

PART 1—SCOPE AND APPLICATION R-3

Section
R101 Scope and General Requirements............ R-3
R102 Alternative Materials, Design and Methods of Construction and Equipment..... R-3

PART 2—ADMINISTRATION AND ENFORCEMENT R-3

R103 Construction Documents R-3
R104 Inspections R-4
R105 Validity................................. R-5
R106 Referenced Standards..................... R-5
R107 Fees.................................... R-5
R108 Stop Work Order R-5
R109 Board of Appeals R-5

CHAPTER 2 DEFINITIONS R-7
Section
R201 General R-7
R202 General Definitions R-7

CHAPTER 3 GENERAL REQUIREMENTS R-11
Section
R301 Climate Zones........................... R-11
R302 Design Conditions R-26
R303 Materials, Systems and Equipment R-26

CHAPTER 4 RESIDENTIAL ENERGY EFFICIENCY R-29
Section
R401 General R-29
R402 Building Thermal Envelope R-29
R403 Systems................................ R-35

R404 Electrical Power and Lighting Systems....... R-37
R405 Simulated Performance Alternative (Performance)......................... R-37
R406 Energy Rating Index Compliance Alternative R-41

CHAPTER 5 EXISTING BUILDINGS......... R-43
Section
R501 General................................ R-43
R502 Additions R-43
R503 Alterations R-44
R504 Repairs............................... R-44
R505 Change of Occupancy or Use.............. R-44

CHAPTER 6 REFERENCED STANDARDS.... R-45

APPENDIX RA RECOMMENDED PROCEDURE FOR WORST-CASE TESTING OF ATMOSPHERIC VENTING SYSTEMS UNDER R402.4 OR R405 CONDITIONS ≤ 5ACH$_{50}$ ... R-49
Section
RA101 Scope R-49
RA201 General Definitions R-49
RA301 Testing Procedure R-49

APPENDIX RB SOLAR-READY PROVISIONS—DETACHED ONE- AND TWO-FAMILY DWELLINGS, MULTIPLE SINGLE-FAMILY DWELLINGS (TOWNHOUSES).. R-51
Section
RB101 Scope R-51
RB102 General Definitions R-51
RB103 Solar-ready Zone R-51

INDEX........ .... R-53

CHAPTER 1 [RE]

SCOPE AND ADMINISTRATION

PART 1—SCOPE AND APPLICATION

SECTION R101
SCOPE AND GENERAL REQUIREMENTS

R101.1 Title. This code shall be known as the *International Energy Conservation Code* of **[NAME OF JURISDICTION]**, and shall be cited as such. It is referred to herein as "this code."

R101.2 Scope. This code applies to *residential buildings* and the building sites and associated systems and equipment.

R101.3 Intent. This code shall regulate the design and construction of buildings for the effective use and conservation of energy over the useful life of each building. This code is intended to provide flexibility to permit the use of innovative approaches and techniques to achieve this objective. This code is not intended to abridge safety, health or environmental requirements contained in other applicable codes or ordinances.

R101.4 Applicability. Where, in any specific case, different sections of this code specify different materials, methods of construction or other requirements, the most restrictive shall govern. Where there is a conflict between a general requirement and a specific requirement, the specific requirement shall govern.

R101.4.1 Mixed occupancy. Where a building includes both *residential* and *commercial* occupancies, each occupancy shall be separately considered and meet the applicable provisions of the IECC—Commercial Provisions or IECC—Residential Provisions.

R101.5 Compliance. *Residential buildings* shall meet the provisions of IECC—Residential Provisions. *Commercial buildings* shall meet the provisions of IECC—Commercial Provisions.

R101.5.1 Compliance materials. The *code official* shall be permitted to approve specific computer software, worksheets, compliance manuals and other similar materials that meet the intent of this code.

SECTION R102
ALTERNATIVE MATERIALS, DESIGN AND METHODS OF CONSTRUCTION AND EQUIPMENT

R102.1 General. The provisions of this code are not intended to prevent the installation of any material or to prohibit any design or method of construction not specifically prescribed by this code, provided that any such alternative has been approved. The code official shall be permitted to approve an alternative material, design or method of construction where the *code official* finds that the proposed design is satisfactory and complies with the intent of the provisions of this code, and that the material, method or work offered is, for the purpose intended, at least the equivalent of that prescribed in this code.

R102.1.1 Above code programs. The *code official* or other authority having jurisdiction shall be permitted to deem a national, state or local energy-efficiency program to exceed the energy efficiency required by this code. Buildings *approved* in writing by such an energy-efficiency program shall be considered in compliance with this code. The requirements identified as "mandatory" in Chapter 4 shall be met.

PART 2—ADMINISTRATION AND ENFORCEMENT

SECTION R103
CONSTRUCTION DOCUMENTS

R103.1 General. Construction documents, technical reports and other supporting data shall be submitted in one or more sets with each application for a permit. The construction documents and technical reports shall be prepared by a registered design professional where required by the statutes of the jurisdiction in which the project is to be constructed. Where special conditions exist, the *code official* is authorized to require necessary construction documents to be prepared by a registered design professional.

Exception: The *code official* is authorized to waive the requirements for construction documents or other supporting data if the *code official* determines they are not necessary to confirm compliance with this code.

R103.2 Information on construction documents. Construction documents shall be drawn to scale upon suitable material. Electronic media documents are permitted to be submitted where *approved* by the *code official*. Construction documents shall be of sufficient clarity to indicate the location, nature and extent of the work proposed, and show in sufficient detail pertinent data and features of the *building*, systems and equipment as herein governed. Details shall include, but are not limited to, the following as applicable:

1. Insulation materials and their *R*-values.
2. Fenestration *U*-factors and solar heat gain coefficients (SHGC).
3. Area-weighted *U*-factor and solar heat gain coefficients (SHGC) calculations.
4. Mechanical system design criteria.
5. Mechanical and service water-heating system and equipment types, sizes and efficiencies.
6. Equipment and system controls.
7. Duct sealing, duct and pipe insulation and location.
8. Air sealing details.

R103.2.1 Building thermal envelope depiction. The *building's thermal envelope* shall be represented on the construction documents.

R103.3 Examination of documents. The *code official* shall examine or cause to be examined the accompanying construction documents and shall ascertain whether the construction indicated and described is in accordance with the requirements of this code and other pertinent laws or ordinances. The *code official* is authorized to utilize a registered design professional, or other *approved* entity not affiliated with the building design or construction, in conducting the review of the plans and specifications for compliance with the code.

R103.3.1 Approval of construction documents. When the *code official* issues a permit where construction documents are required, the construction documents shall be endorsed in writing and stamped "Reviewed for Code Compliance." Such *approved* construction documents shall not be changed, modified or altered without authorization from the *code official*. Work shall be done in accordance with the *approved* construction documents.

One set of construction documents so reviewed shall be retained by the *code official*. The other set shall be returned to the applicant, kept at the site of work and shall be open to inspection by the *code official* or a duly authorized representative.

R103.3.2 Previous approvals. This code shall not require changes in the construction documents, construction or designated occupancy of a structure for which a lawful permit has been heretofore issued or otherwise lawfully authorized, and the construction of which has been pursued in good faith within 180 days after the effective date of this code and has not been abandoned.

R103.3.3 Phased approval. The *code official* shall have the authority to issue a permit for the construction of part of an energy conservation system before the construction documents for the entire system have been submitted or *approved*, provided adequate information and detailed statements have been filed complying with all pertinent requirements of this code. The holders of such permit shall proceed at their own risk without assurance that the permit for the entire energy conservation system will be granted.

R103.4 Amended construction documents. Work shall be installed in accordance with the approved construction documents, and any changes made during construction that are not in compliance with the *approved* construction documents shall be resubmitted for approval as an amended set of construction documents.

R103.5 Retention of construction documents. One set of *approved* construction documents shall be retained by the *code official* for a period of not less than 180 days from date of completion of the permitted work, or as required by state or local laws.

SECTION R104
INSPECTIONS

R104.1 General. Construction or work for which a permit is required shall be subject to inspection by the *code official* or his or her designated agent, and such construction or work shall remain accessible and exposed for inspection purposes until *approved*. It shall be the duty of the permit applicant to cause the work to remain accessible and exposed for inspection purposes. Neither the *code official* nor the jurisdiction shall be liable for expense entailed in the removal or replacement of any material, product, system or building component required to allow inspection to validate compliance with this code.

R104.2 Required inspections. The *code official* or his or her designated agent, upon notification, shall make the inspections set forth in Sections R104.2.1 through R104.2.5.

R104.2.1 Footing and foundation inspection. Inspections associated with footings and foundations shall verify compliance with the code as to *R*-value, location, thickness, depth of burial and protection of insulation as required by the code and *approved* plans and specifications.

R104.2.2 Framing and rough-in inspection. Inspections at framing and rough-in shall be made before application of interior finish and shall verify compliance with the code as to types of insulation and corresponding *R*-values and their correct location and proper installation; fenestration properties (*U*-factor and SHGC) and proper installation; and air leakage controls as required by the code and approved plans and specifications.

R104.2.3 Plumbing rough-in inspection. Inspections at plumbing rough-in shall verify compliance as required by the code and *approved* plans and specifications as to types of insulation and corresponding *R*-values and protection, and required control.

R104.2.4 Mechanical rough-in inspection. Inspections at mechanical rough-in shall verify compliance as required by the code and *approved* plans and specifications as to installed HVAC equipment type and size, required controls, system insulation and corresponding *R*-value, system air leakage control, programmable thermostats, dampers, whole-house ventilation, and minimum fan efficiency.

> **Exception:** Systems serving multiple dwelling units shall be inspected in accordance with Section C104.2.4.

R104.2.5 Final inspection. The *building* shall have a final inspection and shall not be occupied until *approved*. The final inspection shall include verification of the installation of all required *building* systems, equipment and controls and their proper operation and the required number of high-efficacy lamps and fixtures.

R104.3 Reinspection. A *building* shall be reinspected when determined necessary by the *code official*.

R104.4 Approved inspection agencies. The *code official* is authorized to accept reports of third-party inspection agencies not affiliated with the building design or construction, provided such agencies are *approved* as to qualifications and reliability relevant to the building components and systems they are inspecting.

R104.5 Inspection requests. It shall be the duty of the holder of the permit or their duly authorized agent to notify the *code official* when work is ready for inspection. It shall be the duty of the permit holder to provide access to and means for inspections of such work that are required by this code.

R104.6 Reinspection and testing. Where any work or installation does not pass an initial test or inspection, the necessary corrections shall be made to achieve compliance with this code. The work or installation shall then be resubmitted to the *code official* for inspection and testing.

R104.7 Approval. After the prescribed tests and inspections indicate that the work complies in all respects with this code, a notice of approval shall be issued by the *code official*.

R104.7.1 Revocation. The *code official* is authorized to, in writing, suspend or revoke a notice of approval issued under the provisions of this code wherever the certificate is issued in error, or on the basis of incorrect information supplied, or where it is determined that the *building* or structure, premise, or portion thereof is in violation of any ordinance or regulation or any of the provisions of this code.

SECTION R105
VALIDITY

R105.1 General. If a portion of this code is held to be illegal or void, such a decision shall not affect the validity of the remainder of this code.

SECTION R106
REFERENCED STANDARDS

R106.1 Referenced codes and standards. The codes and standards referenced in this code shall be those listed in Chapter 5, and such codes and standards shall be considered as part of the requirements of this code to the prescribed extent of each such reference and as further regulated in Sections R106.1.1 and R106.1.2.

R106.1.1 Conflicts. Where conflicts occur between provisions of this code and referenced codes and standards, the provisions of this code shall apply.

R106.1.2 Provisions in referenced codes and standards. Where the extent of the reference to a referenced code or standard includes subject matter that is within the scope of this code, the provisions of this code, as applicable, shall take precedence over the provisions in the referenced code or standard.

R106.2 Application of references. References to chapter or section numbers, or to provisions not specifically identified by number, shall be construed to refer to such chapter, section or provision of this code.

R106.3 Other laws. The provisions of this code shall not be deemed to nullify any provisions of local, state or federal law.

SECTION R107
FEES

R107.1 Fees. A permit shall not be issued until the fees prescribed in Section R107.2 have been paid, nor shall an amendment to a permit be released until the additional fee, if any, has been paid.

R107.2 Schedule of permit fees. A fee for each permit shall be paid as required, in accordance with the schedule as established by the applicable governing authority.

R107.3 Work commencing before permit issuance. Any person who commences any work before obtaining the necessary permits shall be subject to an additional fee established by the *code official* that shall be in addition to the required permit fees.

R107.4 Related fees. The payment of the fee for the construction, *alteration*, removal or demolition of work done in connection to or concurrently with the work or activity authorized by a permit shall not relieve the applicant or holder of the permit from the payment of other fees that are prescribed by law.

R107.5 Refunds. The *code official* is authorized to establish a refund policy.

SECTION R108
STOP WORK ORDER

R108.1 Authority. Where the *code official* finds any work regulated by this code being performed in a manner either contrary to the provisions of this code or dangerous or unsafe, the *code official* is authorized to issue a stop work order.

R108.2 Issuance. The stop work order shall be in writing and shall be given to the owner of the property involved, to the owner's authorized agent, or to the person doing the work. Upon issuance of a stop work order, the cited work shall immediately cease. The stop work order shall state the reason for the order and the conditions under which the cited work will be permitted to resume.

R108.3 Emergencies. Where an emergency exists, the *code official* shall not be required to give a written notice prior to stopping the work.

R108.4 Failure to comply. Any person who shall continue any work after having been served with a stop work order, except such work as that person is directed to perform to remove a violation or unsafe condition, shall be subject to a fine as set by the applicable governing authority.

SECTION R109
BOARD OF APPEALS

R109.1 General. In order to hear and decide appeals of orders, decisions or determinations made by the *code official* relative to the application and interpretation of this code, there shall be and is hereby created a board of appeals. The *code official* shall be an ex officio member of said board but shall not have a vote on any matter before the board. The board of appeals shall be appointed by the governing body and shall hold office at its pleasure. The board shall adopt rules of procedure for conducting its business, and shall render all decisions and findings in writing to the appellant with a duplicate copy to the *code official*.

R109.2 Limitations on authority. An application for appeal shall be based on a claim that the true intent of this code or the rules legally adopted thereunder have been incorrectly interpreted, the provisions of this code do not fully apply or an equally good or better form of construction is proposed. The board shall not have authority to waive requirements of this code.

R109.3 Qualifications. The board of appeals shall consist of members who are qualified by experience and training and are not employees of the jurisdiction.

CHAPTER 2 [RE]
DEFINITIONS

SECTION R201
GENERAL

R201.1 Scope. Unless stated otherwise, the following words and terms in this code shall have the meanings indicated in this chapter.

R201.2 Interchangeability. Words used in the present tense include the future; words in the masculine gender include the feminine and neuter; the singular number includes the plural and the plural includes the singular.

R201.3 Terms defined in other codes. Terms that are not defined in this code but are defined in the *International Building Code, International Fire Code, International Fuel Gas Code, International Mechanical Code, International Plumbing Code* or the *International Residential Code* shall have the meanings ascribed to them in those codes.

R201.4 Terms not defined. Terms not defined by this chapter shall have ordinarily accepted meanings such as the context implies.

SECTION R202
GENERAL DEFINITIONS

ABOVE-GRADE WALL. A wall more than 50 percent above grade and enclosing *conditioned space*. This includes between-floor spandrels, peripheral edges of floors, roof and basement knee walls, dormer walls, gable end walls, walls enclosing a mansard roof and skylight shafts.

ACCESSIBLE. Admitting close approach as a result of not being guarded by locked doors, elevation or other effective means (see "Readily *accessible*").

ADDITION. An extension or increase in the *conditioned space* floor area or height of a building or structure.

AIR BARRIER. Material(s) assembled and joined together to provide a barrier to air leakage through the building envelope. An air barrier may be a single material or a combination of materials.

ALTERATION. Any construction, retrofit or renovation to an existing structure other than repair or addition that requires a permit. Also, a change in a building, electrical, gas, mechanical or plumbing system that involves an extension, addition or change to the arrangement, type or purpose of the original installation that requires a permit.

APPROVED. Approval by the *code official* as a result of investigation and tests conducted by him or her, or by reason of accepted principles or tests by nationally recognized organizations.

APPROVED AGENCY. An established and recognized agency regularly engaged in conducting tests or furnishing inspection services, when such agency has been approved by the *code official*.

AUTOMATIC. Self-acting, operating by its own mechanism when actuated by some impersonal influence, as, for example, a change in current strength, pressure, temperature or mechanical configuration (see "Manual").

BASEMENT WALL. A wall 50 percent or more below grade and enclosing *conditioned space*.

BUILDING. Any structure used or intended for supporting or sheltering any use or occupancy, including any mechanical systems, service water heating systems and electric power and lighting systems located on the building site and supporting the building.

BUILDING SITE. A contiguous area of land that is under the ownership or control of one entity.

BUILDING THERMAL ENVELOPE. The basement walls, exterior walls, floor, roof and any other building elements that enclose conditioned space or provide a boundary between conditioned space and exempt or unconditioned space.

C-FACTOR (THERMAL CONDUCTANCE). The coefficient of heat transmission (surface to surface) through a building component or assembly, equal to the time rate of heat flow per unit area and the unit temperature difference between the warm side and cold side surfaces $(Btu/h \cdot ft^2 \cdot °F)$ $[W/(m^2 \cdot K)]$.

CIRCULATING HOT WATER SYSTEM. A specifically designed water distribution system where one or more pumps are operated in the service hot water piping to circulate heated water from the water-heating equipment to fixtures and back to the water-heating equipment.

CLIMATE ZONE. A geographical region based on climatic criteria as specified in this code.

CODE OFFICIAL. The officer or other designated authority charged with the administration and enforcement of this code, or a duly authorized representative.

COMMERCIAL BUILDING. For this code, all buildings that are not included in the definition of "Residential building."

CONDITIONED FLOOR AREA. The horizontal projection of the floors associated with the *conditioned space*.

CONDITIONED SPACE. An area, room or space that is enclosed within the building thermal envelope and that is directly or indirectly heated or cooled. Spaces are indirectly heated or cooled where they communicate through openings with conditioned spaces, where they are separated from conditioned spaces by uninsulated walls, floors or ceilings, or where they contain uninsulated ducts, piping or other sources of heating or cooling.

CONTINUOUS AIR BARRIER. A combination of materials and assemblies that restrict or prevent the passage of air through the building thermal envelope.

DEFINITIONS

CONTINUOUS INSULATION (ci). Insulating material that is continuous across all structural members without thermal bridges other than fasteners and service openings. It is installed on the interior or exterior, or is integral to any opaque surface, of the building envelope.

CRAWL SPACE WALL. The opaque portion of a wall that encloses a crawl space and is partially or totally below grade.

CURTAIN WALL. Fenestration products used to create an external nonload-bearing wall that is designed to separate the exterior and interior environments.

DEMAND RECIRCULATION WATER SYSTEM. A water distribution system where pump(s) prime the service hot water piping with heated water upon demand for hot water.

DUCT. A tube or conduit utilized for conveying air. The air passages of self-contained systems are not to be construed as air ducts.

DUCT SYSTEM. A continuous passageway for the transmission of air that, in addition to ducts, includes duct fittings, dampers, plenums, fans and accessory air-handling equipment and appliances.

DWELLING UNIT. A single unit providing complete independent living facilities for one or more persons, including permanent provisions for living, sleeping, eating, cooking and sanitation.

ENERGY ANALYSIS. A method for estimating the annual energy use of the *proposed design* and *standard reference design* based on estimates of energy use.

ENERGY COST. The total estimated annual cost for purchased energy for the building functions regulated by this code, including applicable demand charges.

ENERGY SIMULATION TOOL. An *approved* software program or calculation-based methodology that projects the annual energy use of a building.

ERI REFERENCE DESIGN. A version of the rated design that meets the minimum requirements of the 2006 *International Energy Conservation Code*.

EXTERIOR WALL. Walls including both above-grade walls and basement walls.

FENESTRATION. Products classified as either *vertical fenestration* or *skylights*.

FENESTRATION PRODUCT, SITE-BUILT. A fenestration designed to be made up of field-glazed or field-assembled units using specific factory cut or otherwise factory-formed framing and glazing units. Examples of site-built fenestration include storefront systems, curtain walls and atrium roof systems.

***F*-FACTOR.** The perimeter heat loss factor for slab-on-grade floors (Btu/h·ft·°F) [W/(m·K)].

HEATED SLAB. Slab-on-grade construction in which the heating elements, hydronic tubing, or hot air distribution system is in contact with, or placed within or under, the slab.

HIGH-EFFICACY LAMPS. Compact fluorescent lamps, T-8 or smaller diameter linear fluorescent lamps, or lamps with a minimum efficacy of:

1. 60 lumens per watt for lamps over 40 watts;
2. 50 lumens per watt for lamps over 15 watts to 40 watts; and
3. 40 lumens per watt for lamps 15 watts or less.

HISTORIC BUILDING. Any building or structure that is one or more of the following:

1. Listed, or certified as eligible for listing by the State Historic Preservation Officer or the Keeper of the National Register of Historic Places, in the National Register of Historic Places.
2. Designated as historic under an applicable state or local law.
3. Certified as a contributing resource within a National Register-listed, state-designated or locally designated historic district.

INFILTRATION. The uncontrolled inward air leakage into a building caused by the pressure effects of wind or the effect of differences in the indoor and outdoor air density or both.

INSULATED SIDING. A type of continuous insulation with manufacturer-installed insulating material as an integral part of the cladding product having a minimum *R*-value of R-2.

INSULATING SHEATHING. An insulating board with a core material having a minimum *R*-value of R-2.

LABELED. Equipment, materials or products to which have been affixed a label, seal, symbol or other identifying mark of a nationally recognized testing laboratory, inspection agency or other organization concerned with product evaluation that maintains periodic inspection of the production of the above-labeled items and where labeling indicates either that the equipment, material or product meets identified standards or has been tested and found suitable for a specified purpose.

LISTED. Equipment, materials, products or services included in a list published by an organization acceptable to the *code official* and concerned with evaluation of products or services that maintains periodic inspection of production of *listed* equipment or materials or periodic evaluation of services and where the listing states either that the equipment, material, product or service meets identified standards or has been tested and found suitable for a specified purpose.

LOW-VOLTAGE LIGHTING. Lighting equipment powered through a transformer such as a cable conductor, a rail conductor and track lighting.

MANUAL. Capable of being operated by personal intervention (see "Automatic").

PROPOSED DESIGN. A description of the proposed *building* used to estimate annual energy use for determining compliance based on total building performance.

RATED DESIGN. A description of the proposed *building* used to determine the energy rating index.

DEFINITIONS

READILY ACCESSIBLE. Capable of being reached quickly for operation, renewal or inspection without requiring those to whom ready access is requisite to climb over or remove obstacles or to resort to portable ladders or access equipment (see "*Accessible*").

REPAIR. The reconstruction or renewal of any part of an existing building for the purpose of its maintenance or to correct damage.

REROOFING. The process of recovering or replacing an existing *roof covering*. See "Roof recover" and "Roof replacement."

RESIDENTIAL BUILDING. For this code, includes detached one- and two-family dwellings and multiple single-family dwellings (townhouses) as well as Group R-2, R-3 and R-4 buildings three stories or less in height above grade plane.

ROOF ASSEMBLY. A system designed to provide weather protection and resistance to design loads. The system consists of a roof covering and roof deck or a single component serving as both the roof covering and the roof deck. A roof assembly includes the roof covering, underlayment, roof deck, insulation, vapor retarder and interior finish.

ROOF RECOVER. The process of installing an additional *roof covering* over a prepared existing *roof covering* without removing the existing *roof covering*.

ROOF REPAIR. Reconstruction or renewal of any part of an existing roof for the purposes of its maintenance.

ROOF REPLACEMENT. The process of removing the existing *roof covering*, repairing any damaged substrate and installing a new *roof covering*.

R-VALUE (THERMAL RESISTANCE). The inverse of the time rate of heat flow through a body from one of its bounding surfaces to the other surface for a unit temperature difference between the two surfaces, under steady state conditions, per unit area ($h \cdot ft^2 \cdot °F/Btu$) [$(m^2 \cdot K)/W$].

SERVICE WATER HEATING. Supply of hot water for purposes other than comfort heating.

SKYLIGHT. Glass or other transparent or translucent glazing material installed at a slope of less than 60 degrees (1.05 rad) from horizontal.

SOLAR HEAT GAIN COEFFICIENT (SHGC). The ratio of the solar heat gain entering the space through the fenestration assembly to the incident solar radiation. Solar heat gain includes directly transmitted solar heat and absorbed solar radiation that is then reradiated, conducted or convected into the space.

STANDARD REFERENCE DESIGN. A version of the *proposed design* that meets the minimum requirements of this code and is used to determine the maximum annual energy use requirement for compliance based on total building performance.

SUNROOM. A one-story structure attached to a dwelling with a glazing area in excess of 40 percent of the gross area of the structure's exterior walls and roof.

THERMAL ISOLATION. Physical and space conditioning separation from *conditioned space(s)*. The *conditioned space*(s) shall be controlled as separate zones for heating and cooling or conditioned by separate equipment.

THERMOSTAT. An automatic control device used to maintain temperature at a fixed or adjustable set point.

***U*-FACTOR (THERMAL TRANSMITTANCE).** The coefficient of heat transmission (air to air) through a building component or assembly, equal to the time rate of heat flow per unit area and unit temperature difference between the warm side and cold side air films ($Btu/h \cdot ft^2 \cdot °F$) [$W/(m^2 \cdot K)$].

VENTILATION. The natural or mechanical process of supplying conditioned or unconditioned air to, or removing such air from, any space.

VENTILATION AIR. That portion of supply air that comes from outside (outdoors) plus any recirculated air that has been treated to maintain the desired quality of air within a designated space.

VERTICAL FENESTRATION. Windows (fixed or moveable), opaque doors, glazed doors, glazed block and combination opaque/glazed doors composed of glass or other transparent or translucent glazing materials and installed at a slope of a least 60 degrees (1.05 rad) from horizontal.

VISIBLE TRANSMITTANCE [VT]. The ratio of visible light entering the space through the fenestration product assembly to the incident visible light, Visible Transmittance, includes the effects of glazing material and frame and is expressed as a number between 0 and 1.

WHOLE HOUSE MECHANICAL VENTILATION SYSTEM. An exhaust system, supply system, or combination thereof that is designed to mechanically exchange indoor air with outdoor air when operating continuously or through a programmed intermittent schedule to satisfy the whole house ventilation rates.

ZONE. A space or group of spaces within a building with heating or cooling requirements that are sufficiently similar so that desired conditions can be maintained throughout using a single controlling device.

CHAPTER 3 [RE]
GENERAL REQUIREMENTS

SECTION R301
CLIMATE ZONES

R301.1 General. *Climate zones* from Figure R301.1 or Table R301.1 shall be used in determining the applicable requirements from Chapter 4. Locations not in Table R301.1 (outside the United States) shall be assigned a *climate zone* based on Section R301.3.

R301.2 Warm humid counties. Warm humid counties are identified in Table R301.1 by an asterisk.

R301.3 International climate zones. The *climate zone* for any location outside the United States shall be determined by applying Table R301.3(1) and then Table R301.3(2).

R301.4 Tropical climate zone. The tropical *climate zone* shall be defined as:

1. Hawaii, Puerto Rico, Guam, American Samoa, U.S. Virgin Islands, Commonwealth of Northern Mariana Islands; and

2. Islands in the area between the Tropic of Cancer and the Tropic of Capricorn.

GENERAL REQUIREMENTS

FIGURE R301.1
CLIMATE ZONES

GENERAL REQUIREMENTS

TABLE R301.1
CLIMATE ZONES, MOISTURE REGIMES, AND WARM-HUMID DESIGNATIONS BY STATE, COUNTY AND TERRITORY

Key: A – Moist, B – Dry, C – Marine. Absence of moisture designation indicates moisture regime is irrelevant. Asterisk (*) indicates a warm-humid location.

US STATES

ALABAMA

3A Autauga*
2A Baldwin*
3A Barbour*
3A Bibb
3A Blount
3A Bullock*
3A Butler*
3A Calhoun
3A Chambers
3A Cherokee
3A Chilton
3A Choctaw*
3A Clarke*
3A Clay
3A Cleburne
3A Coffee*
3A Colbert
3A Conecuh*
3A Coosa
3A Covington*
3A Crenshaw*
3A Cullman
3A Dale*
3A Dallas*
3A DeKalb
3A Elmore*
3A Escambia*
3A Etowah
3A Fayette
3A Franklin
3A Geneva*
3A Greene
3A Hale
3A Henry*
3A Houston*
3A Jackson
3A Jefferson
3A Lamar
3A Lauderdale
3A Lawrence
3A Lee
3A Limestone
3A Lowndes*
3A Macon*
3A Madison
3A Marengo*
3A Marion
3A Marshall
2A Mobile*
3A Monroe*
3A Montgomery*
3A Morgan
3A Perry*
3A Pickens
3A Pike*
3A Randolph
3A Russell*
3A Shelby
3A St. Clair
3A Sumter
3A Talladega
3A Tallapoosa
3A Tuscaloosa
3A Walker
3A Washington*
3A Wilcox*
3A Winston

ALASKA

7 Aleutians East
7 Aleutians West
7 Anchorage
8 Bethel
7 Bristol Bay
7 Denali
8 Dillingham
8 Fairbanks North Star
7 Haines
7 Juneau
7 Kenai Peninsula
7 Ketchikan Gateway
7 Kodiak Island
7 Lake and Peninsula
7 Matanuska-Susitna
8 Nome
8 North Slope
8 Northwest Arctic
7 Prince of Wales-Outer Ketchikan
7 Sitka
7 Skagway-Hoonah-Angoon
8 Southeast Fairbanks
7 Valdez-Cordova
8 Wade Hampton
7 Wrangell-Petersburg
7 Yakutat
8 Yukon-Koyukuk

ARIZONA

5B Apache
3B Cochise
5B Coconino
4B Gila
3B Graham
3B Greenlee
2B La Paz
2B Maricopa
3B Mohave
5B Navajo
2B Pima
2B Pinal
3B Santa Cruz
4B Yavapai
2B Yuma

ARKANSAS

3A Arkansas
3A Ashley
4A Baxter
4A Benton
4A Boone
3A Bradley
3A Calhoun
4A Carroll
3A Chicot
3A Clark
3A Clay
3A Cleburne
3A Cleveland
3A Columbia*
3A Conway
3A Craighead
3A Crawford
3A Crittenden
3A Cross
3A Dallas
3A Desha
3A Drew
3A Faulkner
3A Franklin
4A Fulton
3A Garland
3A Grant
3A Greene
3A Hempstead*
3A Hot Spring
3A Howard
3A Independence
4A Izard
3A Jackson
3A Jefferson
3A Johnson
3A Lafayette*
3A Lawrence
3A Lee
3A Lincoln
3A Little River*
3A Logan
3A Lonoke
4A Madison
4A Marion
3A Miller*
3A Mississippi
3A Monroe
3A Montgomery
3A Nevada
4A Newton
3A Ouachita
3A Perry
3A Phillips
3A Pike
3A Poinsett
3A Polk
3A Pope
3A Prairie
3A Pulaski
3A Randolph
3A Saline
3A Scott
4A Searcy
3A Sebastian
3A Sevier*
3A Sharp
3A St. Francis
4A Stone
3A Union*
3A Van Buren
4A Washington
3A White
3A Woodruff
3A Yell

CALIFORNIA

3C Alameda
6B Alpine
4B Amador
3B Butte
4B Calaveras
3B Colusa
3B Contra Costa
4C Del Norte
4B El Dorado
3B Fresno
3B Glenn

(continued)

GENERAL REQUIREMENTS

TABLE R301.1—continued
CLIMATE ZONES, MOISTURE REGIMES, AND WARM-HUMID
DESIGNATIONS BY STATE, COUNTY AND TERRITORY

4C Humboldt
2B Imperial
4B Inyo
3B Kern
3B Kings
4B Lake
5B Lassen
3B Los Angeles
3B Madera
3C Marin
4B Mariposa
3C Mendocino
3B Merced
5B Modoc
6B Mono
3C Monterey
3C Napa
5B Nevada
3B Orange
3B Placer
5B Plumas
3B Riverside
3B Sacramento
3C San Benito
3B San Bernardino
3B San Diego
3C San Francisco
3B San Joaquin
3C San Luis Obispo
3C San Mateo
3C Santa Barbara
3C Santa Clara
3C Santa Cruz
3B Shasta
5B Sierra
5B Siskiyou
3B Solano
3C Sonoma
3B Stanislaus
3B Sutter
3B Tehama
4B Trinity
3B Tulare
4B Tuolumne
3C Ventura
3B Yolo

3B Yuba

COLORADO

5B Adams
6B Alamosa
5B Arapahoe
6B Archuleta
4B Baca
5B Bent
5B Boulder
5B Broomfield
6B Chaffee
5B Cheyenne
7 Clear Creek
6B Conejos
6B Costilla
5B Crowley
6B Custer
5B Delta
5B Denver
6B Dolores
5B Douglas
6B Eagle
5B Elbert
5B El Paso
5B Fremont
5B Garfield
5B Gilpin
7 Grand
7 Gunnison
7 Hinsdale
5B Huerfano
7 Jackson
5B Jefferson
5B Kiowa
5B Kit Carson
7 Lake
5B La Plata
5B Larimer
4B Las Animas
5B Lincoln
5B Logan
5B Mesa
7 Mineral
6B Moffat
5B Montezuma

5B Montrose
5B Morgan
4B Otero
6B Ouray
7 Park
5B Phillips
7 Pitkin
5B Prowers
5B Pueblo
6B Rio Blanco
7 Rio Grande
7 Routt
6B Saguache
7 San Juan
6B San Miguel
5B Sedgwick
7 Summit
5B Teller
5B Washington
5B Weld
5B Yuma

CONNECTICUT

5A (all)

DELAWARE

4A (all)

DISTRICT OF COLUMBIA

4A (all)

FLORIDA

2A Alachua*
2A Baker*
2A Bay*
2A Bradford*
2A Brevard*
1A Broward*
2A Calhoun*
2A Charlotte*
2A Citrus*
2A Clay*
2A Collier*
2A Columbia*
2A DeSoto*
2A Dixie*
2A Duval*

2A Escambia*
2A Flagler*
2A Franklin*
2A Gadsden*
2A Gilchrist*
2A Glades*
2A Gulf*
2A Hamilton*
2A Hardee*
2A Hendry*
2A Hernando*
2A Highlands*
2A Hillsborough*
2A Holmes*
2A Indian River*
2A Jackson*
2A Jefferson*
2A Lafayette*
2A Lake*
2A Lee*
2A Leon*
2A Levy*
2A Liberty*
2A Madison*
2A Manatee*
2A Marion*
2A Martin*
1A Miami-Dade*
1A Monroe*
2A Nassau*
2A Okaloosa*
2A Okeechobee*
2A Orange*
2A Osceola*
2A Palm Beach*
2A Pasco*
2A Pinellas*
2A Polk*
2A Putnam*
2A Santa Rosa*
2A Sarasota*
2A Seminole*
2A St. Johns*
2A St. Lucie*
2A Sumter*
2A Suwannee*

2A Taylor*
2A Union*
2A Volusia*
2A Wakulla*
2A Walton*
2A Washington*

GEORGIA

2A Appling*
2A Atkinson*
2A Bacon*
2A Baker*
3A Baldwin
4A Banks
3A Barrow
3A Bartow
3A Ben Hill*
2A Berrien*
3A Bibb
3A Bleckley*
2A Brantley*
2A Brooks*
2A Bryan*
3A Bulloch*
3A Burke
3A Butts
3A Calhoun*
2A Camden*
3A Candler*
3A Carroll
4A Catoosa
2A Charlton*
2A Chatham*
3A Chattahoochee*
4A Chattooga
3A Cherokee
3A Clarke
3A Clay*
3A Clayton
2A Clinch*
3A Cobb
3A Coffee*
2A Colquitt*
3A Columbia
2A Cook*
3A Coweta

(continued)

GENERAL REQUIREMENTS

**TABLE R301.1—continued
CLIMATE ZONES, MOISTURE REGIMES, AND WARM-HUMID
DESIGNATIONS BY STATE, COUNTY AND TERRITORY**

3A Crawford	2A Lanier*	3A Taylor*	5B Cassia	4A Crawford
3A Crisp*	3A Laurens*	3A Telfair*	6B Clark	5A Cumberland
4A Dade	3A Lee*	3A Terrell*	5B Clearwater	5A DeKalb
4A Dawson	2A Liberty*	2A Thomas*	6B Custer	5A De Witt
2A Decatur*	3A Lincoln	3A Tift*	5B Elmore	5A Douglas
3A DeKalb	2A Long*	2A Toombs*	6B Franklin	5A DuPage
3A Dodge*	2A Lowndes*	4A Towns	6B Fremont	5A Edgar
3A Dooly*	4A Lumpkin	3A Treutlen*	5B Gem	4A Edwards
3A Dougherty*	3A Macon*	3A Troup	5B Gooding	4A Effingham
3A Douglas	3A Madison	3A Turner*	5B Idaho	4A Fayette
3A Early*	3A Marion*	3A Twiggs*	6B Jefferson	5A Ford
2A Echols*	3A McDuffie	4A Union	5B Jerome	4A Franklin
2A Effingham*	2A McIntosh*	3A Upson	5B Kootenai	5A Fulton
3A Elbert	3A Meriwether	4A Walker	5B Latah	4A Gallatin
3A Emanuel*	2A Miller*	3A Walton	6B Lemhi	5A Greene
2A Evans*	2A Mitchell*	2A Ware*	5B Lewis	5A Grundy
4A Fannin	3A Monroe	3A Warren	5B Lincoln	4A Hamilton
3A Fayette	3A Montgomery*	3A Washington	6B Madison	5A Hancock
4A Floyd	3A Morgan	2A Wayne*	5B Minidoka	4A Hardin
3A Forsyth	4A Murray	3A Webster*	5B Nez Perce	5A Henderson
4A Franklin	3A Muscogee	3A Wheeler*	6B Oneida	5A Henry
3A Fulton	3A Newton	4A White	5B Owyhee	5A Iroquois
4A Gilmer	3A Oconee	4A Whitfield	5B Payette	4A Jackson
3A Glascock	3A Oglethorpe	3A Wilcox*	5B Power	4A Jasper
2A Glynn*	3A Paulding	3A Wilkes	5B Shoshone	4A Jefferson
4A Gordon	3A Peach*	3A Wilkinson	6B Teton	5A Jersey
2A Grady*	4A Pickens	3A Worth*	5B Twin Falls	5A Jo Daviess
3A Greene	2A Pierce*	**HAWAII**	6B Valley	4A Johnson
3A Gwinnett	3A Pike	1A (all)*	5B Washington	5A Kane
4A Habersham	3A Polk	**IDAHO**	**ILLINOIS**	5A Kankakee
4A Hall	3A Pulaski*	5B Ada	5A Adams	5A Kendall
3A Hancock	3A Putnam	6B Adams	4A Alexander	5A Knox
3A Haralson	3A Quitman*	6B Bannock	4A Bond	5A Lake
3A Harris	4A Rabun	6B Bear Lake	5A Boone	5A La Salle
3A Hart	3A Randolph*	5B Benewah	5A Brown	4A Lawrence
3A Heard	3A Richmond	6B Bingham	5A Bureau	5A Lee
3A Henry	3A Rockdale	6B Blaine	5A Calhoun	5A Livingston
3A Houston*	3A Schley*	6B Boise	5A Carroll	5A Logan
3A Irwin*	3A Screven*	6B Bonner	5A Cass	5A Macon
3A Jackson	2A Seminole*	6B Bonneville	5A Champaign	4A Macoupin
3A Jasper	3A Spalding	6B Boundary	4A Christian	4A Madison
2A Jeff Davis*	4A Stephens	6B Butte	5A Clark	4A Marion
3A Jefferson	3A Stewart*	6B Camas	4A Clay	5A Marshall
3A Jenkins*	3A Sumter*	5B Canyon	4A Clinton	5A Mason
3A Johnson*	3A Talbot	6B Caribou	5A Coles	4A Massac
3A Jones	3A Taliaferro		5A Cook	5A McDonough
3A Lamar	2A Tattnall*			5A McHenry

(continued)

GENERAL REQUIREMENTS

TABLE R301.1—continued
CLIMATE ZONES, MOISTURE REGIMES, AND WARM-HUMID DESIGNATIONS BY STATE, COUNTY AND TERRITORY

5A McLean	5A Boone	5A Miami	5A Appanoose	5A Jasper
5A Menard	4A Brown	4A Monroe	5A Audubon	5A Jefferson
5A Mercer	5A Carroll	5A Montgomery	5A Benton	5A Johnson
4A Monroe	5A Cass	5A Morgan	6A Black Hawk	5A Jones
4A Montgomery	4A Clark	5A Newton	5A Boone	5A Keokuk
5A Morgan	5A Clay	5A Noble	6A Bremer	6A Kossuth
5A Moultrie	5A Clinton	4A Ohio	6A Buchanan	5A Lee
5A Ogle	4A Crawford	4A Orange	6A Buena Vista	5A Linn
5A Peoria	4A Daviess	5A Owen	6A Butler	5A Louisa
4A Perry	4A Dearborn	5A Parke	6A Calhoun	5A Lucas
5A Piatt	5A Decatur	4A Perry	5A Carroll	6A Lyon
5A Pike	5A De Kalb	4A Pike	5A Cass	5A Madison
4A Pope	5A Delaware	5A Porter	5A Cedar	5A Mahaska
4A Pulaski	4A Dubois	4A Posey	6A Cerro Gordo	5A Marion
5A Putnam	5A Elkhart	5A Pulaski	6A Cherokee	5A Marshall
4A Randolph	5A Fayette	5A Putnam	6A Chickasaw	5A Mills
4A Richland	4A Floyd	5A Randolph	5A Clarke	6A Mitchell
5A Rock Island	5A Fountain	4A Ripley	6A Clay	5A Monona
4A Saline	5A Franklin	5A Rush	6A Clayton	5A Monroe
5A Sangamon	5A Fulton	4A Scott	5A Clinton	5A Montgomery
5A Schuyler	4A Gibson	5A Shelby	5A Crawford	5A Muscatine
5A Scott	5A Grant	4A Spencer	5A Dallas	6A O'Brien
4A Shelby	4A Greene	5A Starke	5A Davis	6A Osceola
5A Stark	5A Hamilton	5A Steuben	5A Decatur	5A Page
4A St. Clair	5A Hancock	5A St. Joseph	6A Delaware	6A Palo Alto
5A Stephenson	4A Harrison	4A Sullivan	5A Des Moines	6A Plymouth
5A Tazewell	5A Hendricks	4A Switzerland	6A Dickinson	6A Pocahontas
4A Union	5A Henry	5A Tippecanoe	5A Dubuque	5A Polk
5A Vermilion	5A Howard	5A Tipton	6A Emmet	5A Pottawattamie
4A Wabash	5A Huntington	5A Union	6A Fayette	5A Poweshiek
5A Warren	4A Jackson	4A Vanderburgh	6A Floyd	5A Ringgold
4A Washington	5A Jasper	5A Vermillion	6A Franklin	6A Sac
4A Wayne	5A Jay	5A Vigo	5A Fremont	5A Scott
4A White	4A Jefferson	5A Wabash	5A Greene	5A Shelby
5A Whiteside	4A Jennings	5A Warren	6A Grundy	6A Sioux
5A Will	5A Johnson	4A Warrick	5A Guthrie	5A Story
4A Williamson	4A Knox	4A Washington	6A Hamilton	5A Tama
5A Winnebago	5A Kosciusko	5A Wayne	6A Hancock	5A Taylor
5A Woodford	5A LaGrange	5A Wells	6A Hardin	5A Union
INDIANA	5A Lake	5A White	5A Harrison	5A Van Buren
	5A LaPorte	5A Whitley	5A Henry	5A Wapello
5A Adams	4A Lawrence	**IOWA**	6A Howard	5A Warren
5A Allen	5A Madison		6A Humboldt	5A Washington
5A Bartholomew	5A Marion	5A Adair	6A Ida	5A Wayne
5A Benton	5A Marshall	5A Adams	5A Iowa	6A Webster
5A Blackford	4A Martin	6A Allamakee	5A Jackson	6A Winnebago

(continued)

GENERAL REQUIREMENTS

TABLE R301.1—continued
CLIMATE ZONES, MOISTURE REGIMES, AND WARM-HUMID
DESIGNATIONS BY STATE, COUNTY AND TERRITORY

6A Winneshiek
5A Woodbury
6A Worth
6A Wright

KANSAS

4A Allen
4A Anderson
4A Atchison
4A Barber
4A Barton
4A Bourbon
4A Brown
4A Butler
4A Chase
4A Chautauqua
4A Cherokee
5A Cheyenne
4A Clark
4A Clay
5A Cloud
4A Coffey
4A Comanche
4A Cowley
4A Crawford
5A Decatur
4A Dickinson
4A Doniphan
4A Douglas
4A Edwards
4A Elk
5A Ellis
4A Ellsworth
4A Finney
4A Ford
4A Franklin
4A Geary
5A Gove
5A Graham
4A Grant
4A Gray
5A Greeley
4A Greenwood
5A Hamilton
4A Harper
4A Harvey

4A Haskell
4A Hodgeman
4A Jackson
4A Jefferson
5A Jewell
4A Johnson
4A Kearny
4A Kingman
4A Kiowa
4A Labette
5A Lane
4A Leavenworth
4A Lincoln
4A Linn
5A Logan
4A Lyon
4A Marion
4A Marshall
4A McPherson
4A Meade
4A Miami
5A Mitchell
4A Montgomery
4A Morris
4A Morton
4A Nemaha
4A Neosho
5A Ness
5A Norton
4A Osage
5A Osborne
4A Ottawa
4A Pawnee
5A Phillips
4A Pottawatomie
4A Pratt
5A Rawlins
4A Reno
5A Republic
4A Rice
4A Riley
5A Rooks
4A Rush
4A Russell
4A Saline
5A Scott

4A Sedgwick
4A Seward
4A Shawnee
5A Sheridan
5A Sherman
5A Smith
4A Stafford
4A Stanton
4A Stevens
4A Sumner
5A Thomas
5A Trego
4A Wabaunsee
5A Wallace
4A Washington
5A Wichita
4A Wilson
4A Woodson
4A Wyandotte

KENTUCKY

4A (all)

LOUISIANA

2A Acadia*
2A Allen*
2A Ascension*
2A Assumption*
2A Avoyelles*
2A Beauregard*
3A Bienville*
3A Bossier*
3A Caddo*
2A Calcasieu*
3A Caldwell*
2A Cameron*
3A Catahoula*
3A Claiborne*
3A Concordia*
3A De Soto*
2A East Baton Rouge*
3A East Carroll
2A East Feliciana*
2A Evangeline*
3A Franklin*
3A Grant*
2A Iberia*

2A Iberville*
3A Jackson*
2A Jefferson*
2A Jefferson Davis*
2A Lafayette*
2A Lafourche*
3A La Salle*
3A Lincoln*
2A Livingston*
3A Madison*
3A Morehouse
3A Natchitoches*
2A Orleans*
3A Ouachita*
2A Plaquemines*
2A Pointe Coupee*
2A Rapides*
3A Red River*
3A Richland*
3A Sabine*
2A St. Bernard*
2A St. Charles*
2A St. Helena*
2A St. James*
2A St. John the Baptist*
2A St. Landry*
2A St. Martin*
2A St. Mary*
2A St. Tammany*
2A Tangipahoa*
3A Tensas*
2A Terrebonne*
3A Union*
2A Vermilion*
3A Vernon*
2A Washington*
3A Webster*
2A West Baton Rouge*
3A West Carroll
2A West Feliciana*
3A Winn*

MAINE

6A Androscoggin
7 Aroostook

6A Cumberland
6A Franklin
6A Hancock
6A Kennebec
6A Knox
6A Lincoln
6A Oxford
6A Penobscot
6A Piscataquis
6A Sagadahoc
6A Somerset
6A Waldo
6A Washington
6A York

MARYLAND

4A Allegany
4A Anne Arundel
4A Baltimore
4A Baltimore (city)
4A Calvert
4A Caroline
4A Carroll
4A Cecil
4A Charles
4A Dorchester
4A Frederick
5A Garrett
4A Harford
4A Howard
4A Kent
4A Montgomery
4A Prince George's
4A Queen Anne's
4A Somerset
4A St. Mary's
4A Talbot
4A Washington
4A Wicomico
4A Worcester

MASSACHSETTS

5A (all)

MICHIGAN

6A Alcona
6A Alger

(continued)

GENERAL REQUIREMENTS

TABLE R301.1—continued
CLIMATE ZONES, MOISTURE REGIMES, AND WARM-HUMID DESIGNATIONS BY STATE, COUNTY AND TERRITORY

5A Allegan	7 Mackinac	6A Carver	7 Otter Tail	3A Clarke
6A Alpena	5A Macomb	7 Cass	7 Pennington	3A Clay
6A Antrim	6A Manistee	6A Chippewa	7 Pine	3A Coahoma
6A Arenac	6A Marquette	6A Chisago	6A Pipestone	3A Copiah*
7 Baraga	6A Mason	7 Clay	7 Polk	3A Covington*
5A Barry	6A Mecosta	7 Clearwater	6A Pope	3A DeSoto
5A Bay	6A Menominee	7 Cook	6A Ramsey	3A Forrest*
6A Benzie	5A Midland	6A Cottonwood	7 Red Lake	3A Franklin*
5A Berrien	6A Missaukee	7 Crow Wing	6A Redwood	3A George*
5A Branch	5A Monroe	6A Dakota	6A Renville	3A Greene*
5A Calhoun	5A Montcalm	6A Dodge	6A Rice	3A Grenada
5A Cass	6A Montmorency	6A Douglas	6A Rock	2A Hancock*
6A Charlevoix	5A Muskegon	6A Faribault	7 Roseau	2A Harrison*
6A Cheboygan	6A Newaygo	6A Fillmore	6A Scott	3A Hinds*
7 Chippewa	5A Oakland	6A Freeborn	6A Sherburne	3A Holmes
6A Clare	6A Oceana	6A Goodhue	6A Sibley	3A Humphreys
5A Clinton	6A Ogemaw	7 Grant	6A Stearns	3A Issaquena
6A Crawford	7 Ontonagon	6A Hennepin	6A Steele	3A Itawamba
6A Delta	6A Osceola	6A Houston	6A Stevens	2A Jackson*
6A Dickinson	6A Oscoda	7 Hubbard	7 St. Louis	3A Jasper
5A Eaton	6A Otsego	6A Isanti	6A Swift	3A Jefferson*
6A Emmet	5A Ottawa	7 Itasca	6A Todd	3A Jefferson Davis*
5A Genesee	6A Presque Isle	6A Jackson	6A Traverse	3A Jones*
6A Gladwin	6A Roscommon	7 Kanabec	6A Wabasha	3A Kemper
7 Gogebic	5A Saginaw	6A Kandiyohi	7 Wadena	3A Lafayette
6A Grand Traverse	6A Sanilac	7 Kittson	6A Waseca	3A Lamar*
5A Gratiot	7 Schoolcraft	7 Koochiching	6A Washington	3A Lauderdale
5A Hillsdale	5A Shiawassee	6A Lac qui Parle	6A Watonwan	3A Lawrence*
7 Houghton	5A St. Clair	7 Lake	7 Wilkin	3A Leake
6A Huron	5A St. Joseph	7 Lake of the Woods	6A Winona	3A Lee
5A Ingham	5A Tuscola	6A Le Sueur	6A Wright	3A Leflore
5A Ionia	5A Van Buren	6A Lincoln	6A Yellow Medicine	3A Lincoln*
6A Iosco	5A Washtenaw	6A Lyon		3A Lowndes
7 Iron	5A Wayne	7 Mahnomen	**MISSISSIPPI**	3A Madison
6A Isabella	6A Wexford	7 Marshall	3A Adams*	3A Marion*
5A Jackson	**MINNESOTA**	6A Martin	3A Alcorn	3A Marshall
5A Kalamazoo	7 Aitkin	6A McLeod	3A Amite*	3A Monroe
6A Kalkaska	6A Anoka	6A Meeker	3A Attala	3A Montgomery
5A Kent	7 Becker	7 Mille Lacs	3A Benton	3A Neshoba
7 Keweenaw	7 Beltrami	6A Morrison	3A Bolivar	3A Newton
6A Lake	6A Benton	6A Mower	3A Calhoun	3A Noxubee
5A Lapeer	6A Big Stone	6A Murray	3A Carroll	3A Oktibbeha
6A Leelanau	6A Blue Earth	6A Nicollet	3A Chickasaw	3A Panola
5A Lenawee	6A Brown	6A Nobles	3A Choctaw	2A Pearl River*
5A Livingston	7 Carlton	7 Norman	3A Claiborne*	3A Perry*
7 Luce		6A Olmsted		3A Pike*

(continued)

GENERAL REQUIREMENTS

TABLE R301.1—continued
CLIMATE ZONES, MOISTURE REGIMES, AND WARM-HUMID DESIGNATIONS BY STATE, COUNTY AND TERRITORY

3A Pontotoc	5A Chariton	4A Mississippi	4A Webster	4A Cumberland
3A Prentiss	4A Christian	4A Moniteau	5A Worth	4A Essex
3A Quitman	5A Clark	4A Monroe	4A Wright	4A Gloucester
3A Rankin*	4A Clay	4A Montgomery	**MONTANA**	4A Hudson
3A Scott	5A Clinton	4A Morgan	6B (all)	5A Hunterdon
3A Sharkey	4A Cole	4A New Madrid	**NEBRASKA**	5A Mercer
3A Simpson*	4A Cooper	4A Newton	5A (all)	4A Middlesex
3A Smith*	4A Crawford	5A Nodaway	**NEVADA**	4A Monmouth
2A Stone*	4A Dade	4A Oregon	5B Carson City (city)	5A Morris
3A Sunflower	4A Dallas	4A Osage	5B Churchill	4A Ocean
3A Tallahatchie	5A Daviess	4A Ozark	3B Clark	5A Passaic
3A Tate	5A DeKalb	4A Pemiscot	5B Douglas	4A Salem
3A Tippah	4A Dent	4A Perry	5B Elko	5A Somerset
3A Tishomingo	4A Douglas	4A Pettis	5B Esmeralda	5A Sussex
3A Tunica	4A Dunklin	4A Phelps	5B Eureka	4A Union
3A Union	4A Franklin	5A Pike	5B Humboldt	5A Warren
3A Walthall*	4A Gasconade	4A Platte	5B Lander	**NEW MEXICO**
3A Warren*	5A Gentry	4A Polk	5B Lincoln	4B Bernalillo
3A Washington	4A Greene	4A Pulaski	5B Lyon	5B Catron
3A Wayne*	5A Grundy	5A Putnam	5B Mineral	3B Chaves
3A Webster	5A Harrison	5A Ralls	5B Nye	4B Cibola
3A Wilkinson*	4A Henry	4A Randolph	5B Pershing	5B Colfax
3A Winston	4A Hickory	4A Ray	5B Storey	4B Curry
3A Yalobusha	5A Holt	4A Reynolds	5B Washoe	4B DeBaca
3A Yazoo	4A Howard	4A Ripley	5B White Pine	3B Dona Ana
MISSOURI	4A Howell	4A Saline	**NEW HAMPSHIRE**	3B Eddy
5A Adair	4A Iron	5A Schuyler	6A Belknap	4B Grant
5A Andrew	4A Jackson	5A Scotland	6A Carroll	4B Guadalupe
5A Atchison	4A Jasper	4A Scott	5A Cheshire	5B Harding
4A Audrain	4A Jefferson	4A Shannon	6A Coos	3B Hidalgo
4A Barry	4A Johnson	5A Shelby	6A Grafton	3B Lea
4A Barton	5A Knox	4A St. Charles	5A Hillsborough	4B Lincoln
4A Bates	4A Laclede	4A St. Clair	6A Merrimack	5B Los Alamos
4A Benton	4A Lafayette	4A St. Francois	5A Rockingham	3B Luna
4A Bollinger	4A Lawrence	4A St. Louis	5A Strafford	5B McKinley
4A Boone	5A Lewis	4A St. Louis (city)	6A Sullivan	5B Mora
5A Buchanan	4A Lincoln	4A Ste. Genevieve	**NEW JERSEY**	3B Otero
4A Butler	5A Linn	4A Stoddard	4A Atlantic	4B Quay
5A Caldwell	5A Livingston	4A Stone	5A Bergen	5B Rio Arriba
4A Callaway	5A Macon	5A Sullivan	4A Burlington	4B Roosevelt
4A Camden	4A Madison	4A Taney	4A Camden	5B Sandoval
4A Cape Girardeau	4A Maries	4A Texas	4A Cape May	5B San Juan
4A Carroll	5A Marion	4A Vernon		5B San Miguel
4A Carter	4A McDonald	4A Warren		5B Santa Fe
4A Cass	5A Mercer	4A Washington		4B Sierra
4A Cedar	4A Miller	4A Wayne		4B Socorro

(continued)

GENERAL REQUIREMENTS

TABLE R301.1—continued
CLIMATE ZONES, MOISTURE REGIMES, AND WARM-HUMID
DESIGNATIONS BY STATE, COUNTY AND TERRITORY

5B Taos	4A Queens	4A Clay	4A Orange	7 Divide
5B Torrance	5A Rensselaer	4A Cleveland	3A Pamlico	6A Dunn
4B Union	4A Richmond	3A Columbus*	3A Pasquotank	7 Eddy
4B Valencia	5A Rockland	3A Craven	3A Pender*	6A Emmons
NEW YORK	5A Saratoga	3A Cumberland	3A Perquimans	7 Foster
	5A Schenectady	3A Currituck	4A Person	6A Golden Valley
5A Albany	6A Schoharie	3A Dare	3A Pitt	7 Grand Forks
6A Allegany	6A Schuyler	3A Davidson	4A Polk	6A Grant
4A Bronx	5A Seneca	4A Davie	3A Randolph	7 Griggs
6A Broome	6A Steuben	3A Duplin	3A Richmond	6A Hettinger
6A Cattaraugus	6A St. Lawrence	4A Durham	3A Robeson	7 Kidder
5A Cayuga	4A Suffolk	3A Edgecombe	4A Rockingham	6A LaMoure
5A Chautauqua	6A Sullivan	4A Forsyth	3A Rowan	6A Logan
5A Chemung	5A Tioga	4A Franklin	4A Rutherford	7 McHenry
6A Chenango	6A Tompkins	3A Gaston	3A Sampson	6A McIntosh
6A Clinton	6A Ulster	4A Gates	3A Scotland	6A McKenzie
5A Columbia	6A Warren	4A Graham	3A Stanly	7 McLean
5A Cortland	5A Washington	4A Granville	4A Stokes	6A Mercer
6A Delaware	5A Wayne	3A Greene	4A Surry	6A Morton
5A Dutchess	4A Westchester	4A Guilford	4A Swain	7 Mountrail
5A Erie	6A Wyoming	4A Halifax	4A Transylvania	7 Nelson
6A Essex	5A Yates	4A Harnett	3A Tyrrell	6A Oliver
6A Franklin	**NORTH CAROLINA**	4A Haywood	3A Union	7 Pembina
6A Fulton		4A Henderson	4A Vance	7 Pierce
5A Genesee		4A Hertford	4A Wake	7 Ramsey
5A Greene	4A Alamance	3A Hoke	4A Warren	6A Ransom
6A Hamilton	4A Alexander	3A Hyde	3A Washington	7 Renville
6A Herkimer	5A Alleghany	4A Iredell	5A Watauga	6A Richland
6A Jefferson	3A Anson	4A Jackson	3A Wayne	7 Rolette
4A Kings	5A Ashe	3A Johnston	4A Wilkes	6A Sargent
6A Lewis	5A Avery	3A Jones	3A Wilson	7 Sheridan
5A Livingston	3A Beaufort	4A Lee	4A Yadkin	6A Sioux
6A Madison	4A Bertie	3A Lenoir	5A Yancey	6A Slope
5A Monroe	3A Bladen	4A Lincoln	**NORTH DAKOTA**	6A Stark
6A Montgomery	3A Brunswick*	4A Macon		7 Steele
4A Nassau	4A Buncombe	4A Madison	6A Adams	7 Stutsman
4A New York	4A Burke	3A Martin	7 Barnes	7 Towner
5A Niagara	3A Cabarrus	4A McDowell	7 Benson	7 Traill
6A Oneida	4A Caldwell	3A Mecklenburg	6A Billings	7 Walsh
5A Onondaga	3A Camden	5A Mitchell	7 Bottineau	7 Ward
5A Ontario	3A Carteret*	3A Montgomery	6A Bowman	7 Wells
5A Orange	4A Caswell	3A Moore	7 Burke	7 Williams
5A Orleans	4A Catawba	4A Nash	6A Burleigh	**OHIO**
5A Oswego	4A Chatham	3A New Hanover*	7 Cass	
6A Otsego	4A Cherokee	4A Northampton	7 Cavalier	4A Adams
5A Putnam	3A Chowan	3A Onslow*	6A Dickey	5A Allen

(continued)

GENERAL REQUIREMENTS

TABLE R301.1—continued
CLIMATE ZONES, MOISTURE REGIMES, AND WARM-HUMID
DESIGNATIONS BY STATE, COUNTY AND TERRITORY

5A Ashland	5A Mahoning	3A Bryan	3A Okfuskee	4C Linn
5A Ashtabula	5A Marion	3A Caddo	3A Oklahoma	5B Malheur
5A Athens	5A Medina	3A Canadian	3A Okmulgee	4C Marion
5A Auglaize	5A Meigs	3A Carter	3A Osage	5B Morrow
5A Belmont	5A Mercer	3A Cherokee	3A Ottawa	4C Multnomah
4A Brown	5A Miami	3A Choctaw	3A Pawnee	4C Polk
5A Butler	5A Monroe	4B Cimarron	3A Payne	5B Sherman
5A Carroll	5A Montgomery	3A Cleveland	3A Pittsburg	4C Tillamook
5A Champaign	5A Morgan	3A Coal	3A Pontotoc	5B Umatilla
5A Clark	5A Morrow	3A Comanche	3A Pottawatomie	5B Union
4A Clermont	5A Muskingum	3A Cotton	3A Pushmataha	5B Wallowa
5A Clinton	5A Noble	3A Craig	3A Roger Mills	5B Wasco
5A Columbiana	5A Ottawa	3A Creek	3A Rogers	4C Washington
5A Coshocton	5A Paulding	3A Custer	3A Seminole	5B Wheeler
5A Crawford	5A Perry	3A Delaware	3A Sequoyah	4C Yamhill
5A Cuyahoga	5A Pickaway	3A Dewey	3A Stephens	**PENNSYLVANIA**
5A Darke	4A Pike	3A Ellis	4B Texas	5A Adams
5A Defiance	5A Portage	3A Garfield	3A Tillman	5A Allegheny
5A Delaware	5A Preble	3A Garvin	3A Tulsa	5A Armstrong
5A Erie	5A Putnam	3A Grady	3A Wagoner	5A Beaver
5A Fairfield	5A Richland	3A Grant	3A Washington	5A Bedford
5A Fayette	5A Ross	3A Greer	3A Washita	5A Berks
5A Franklin	5A Sandusky	3A Harmon	3A Woods	5A Blair
5A Fulton	4A Scioto	3A Harper	3A Woodward	5A Bradford
4A Gallia	5A Seneca	3A Haskell	**OREGON**	4A Bucks
5A Geauga	5A Shelby	3A Hughes	5B Baker	5A Butler
5A Greene	5A Stark	3A Jackson	4C Benton	5A Cambria
5A Guernsey	5A Summit	3A Jefferson	4C Clackamas	6A Cameron
4A Hamilton	5A Trumbull	3A Johnston	4C Clatsop	5A Carbon
5A Hancock	5A Tuscarawas	3A Kay	4C Columbia	5A Centre
5A Hardin	5A Union	3A Kingfisher	4C Coos	4A Chester
5A Harrison	5A Van Wert	3A Kiowa	5B Crook	5A Clarion
5A Henry	5A Vinton	3A Latimer	4C Curry	6A Clearfield
5A Highland	5A Warren	3A Le Flore	5B Deschutes	5A Clinton
5A Hocking	4A Washington	3A Lincoln	4C Douglas	5A Columbia
5A Holmes	5A Wayne	3A Logan	5B Gilliam	5A Crawford
5A Huron	5A Williams	3A Love	5B Grant	5A Cumberland
5A Jackson	5A Wood	3A Major	5B Harney	5A Dauphin
5A Jefferson	5A Wyandot	3A Marshall	5B Hood River	4A Delaware
5A Knox	**OKLAHOMA**	3A Mayes	4C Jackson	6A Elk
5A Lake	3A Adair	3A McClain	5B Jefferson	5A Erie
4A Lawrence	3A Alfalfa	3A McCurtain	4C Josephine	5A Fayette
5A Licking	3A Atoka	3A McIntosh	5B Klamath	5A Forest
5A Logan	4B Beaver	3A Murray	5B Lake	5A Franklin
5A Lorain	3A Beckham	3A Muskogee	4C Lane	5A Fulton
5A Lucas	3A Blaine	3A Noble	4C Lincoln	5A Greene
5A Madison		3A Nowata		

(continued)

GENERAL REQUIREMENTS

TABLE R301.1—continued
CLIMATE ZONES, MOISTURE REGIMES, AND WARM-HUMID DESIGNATIONS BY STATE, COUNTY AND TERRITORY

5A Huntingdon
5A Indiana
5A Jefferson
5A Juniata
5A Lackawanna
5A Lancaster
5A Lawrence
5A Lebanon
5A Lehigh
5A Luzerne
5A Lycoming
6A McKean
5A Mercer
5A Mifflin
5A Monroe
4A Montgomery
5A Montour
5A Northampton
5A Northumberland
5A Perry
4A Philadelphia
5A Pike
6A Potter
5A Schuylkill
5A Snyder
5A Somerset
5A Sullivan
6A Susquehanna
6A Tioga
5A Union
5A Venango
5A Warren
5A Washington
6A Wayne
5A Westmoreland
5A Wyoming
4A York

RHODE ISLAND

5A (all)

SOUTH CAROLINA

3A Abbeville
3A Aiken
3A Allendale*
3A Anderson

3A Bamberg*
3A Barnwell*
3A Beaufort*
3A Berkeley*
3A Calhoun
3A Charleston*
3A Cherokee
3A Chester
3A Chesterfield
3A Clarendon
3A Colleton*
3A Darlington
3A Dillon
3A Dorchester*
3A Edgefield
3A Fairfield
3A Florence
3A Georgetown*
3A Greenville
3A Greenwood
3A Hampton*
3A Horry*
3A Jasper*
3A Kershaw
3A Lancaster
3A Laurens
3A Lee
3A Lexington
3A Marion
3A Marlboro
3A McCormick
3A Newberry
3A Oconee
3A Orangeburg
3A Pickens
3A Richland
3A Saluda
3A Spartanburg
3A Sumter
3A Union
3A Williamsburg
3A York

SOUTH DAKOTA

6A Aurora
6A Beadle

5A Bennett
5A Bon Homme
6A Brookings
6A Brown
6A Brule
6A Buffalo
6A Butte
6A Campbell
5A Charles Mix
6A Clark
5A Clay
6A Codington
6A Corson
6A Custer
6A Davison
6A Day
6A Deuel
6A Dewey
5A Douglas
6A Edmunds
6A Fall River
6A Faulk
6A Grant
5A Gregory
6A Haakon
6A Hamlin
6A Hand
6A Hanson
6A Harding
6A Hughes
5A Hutchinson
6A Hyde
5A Jackson
6A Jerauld
6A Jones
6A Kingsbury
6A Lake
6A Lawrence
6A Lincoln
6A Lyman
6A Marshall
6A McCook
6A McPherson
6A Meade
5A Mellette
6A Miner

6A Minnehaha
6A Moody
6A Pennington
6A Perkins
6A Potter
6A Roberts
6A Sanborn
6A Shannon
6A Spink
6A Stanley
6A Sully
5A Todd
5A Tripp
6A Turner
5A Union
6A Walworth
5A Yankton
6A Ziebach

TENNESSEE

4A Anderson
4A Bedford
4A Benton
4A Bledsoe
4A Blount
4A Bradley
4A Campbell
4A Cannon
4A Carroll
4A Carter
4A Cheatham
3A Chester
4A Claiborne
4A Clay
4A Cocke
4A Coffee
3A Crockett
4A Cumberland
4A Davidson
4A Decatur
4A DeKalb
4A Dickson
3A Dyer
3A Fayette
4A Fentress
4A Franklin

4A Gibson
4A Giles
4A Grainger
4A Greene
4A Grundy
4A Hamblen
4A Hamilton
4A Hancock
3A Hardeman
3A Hardin
4A Hawkins
3A Haywood
3A Henderson
4A Henry
4A Hickman
4A Houston
4A Humphreys
4A Jackson
4A Jefferson
4A Johnson
4A Knox
3A Lake
3A Lauderdale
4A Lawrence
4A Lewis
4A Lincoln
4A Loudon
4A Macon
3A Madison
4A Marion
4A Marshall
4A Maury
4A McMinn
3A McNairy
4A Meigs
4A Monroe
4A Montgomery
4A Moore
4A Morgan
4A Obion
4A Overton
4A Perry
4A Pickett
4A Polk
4A Putnam
4A Rhea

(continued)

GENERAL REQUIREMENTS

TABLE R301.1—continued
CLIMATE ZONES, MOISTURE REGIMES, AND WARM-HUMID DESIGNATIONS BY STATE, COUNTY AND TERRITORY

4A Roane	3B Brewster	3B Ector	3B Howard	3B McCulloch
4A Robertson	4B Briscoe	2B Edwards	3B Hudspeth	2A McLennan*
4A Rutherford	2A Brooks*	3A Ellis*	3A Hunt*	2A McMullen*
4A Scott	3A Brown*	3B El Paso	4B Hutchinson	2B Medina
4A Sequatchie	2A Burleson*	3A Erath*	3B Irion	3B Menard
4A Sevier	3A Burnet*	2A Falls*	3A Jack	3B Midland
3A Shelby	2A Caldwell*	3A Fannin	2A Jackson*	2A Milam*
4A Smith	2A Calhoun*	2A Fayette*	2A Jasper*	3A Mills*
4A Stewart	3B Callahan	3B Fisher	3B Jeff Davis	3B Mitchell
4A Sullivan	2A Cameron*	4B Floyd	2A Jefferson*	3A Montague
4A Sumner	3A Camp*	3B Foard	2A Jim Hogg*	2A Montgomery*
3A Tipton	4B Carson	2A Fort Bend*	2A Jim Wells*	4B Moore
4A Trousdale	3A Cass*	3A Franklin*	2A Johnson*	3A Morris*
4A Unicoi	4B Castro	2A Freestone*	3B Jones	3B Motley
4A Union	2A Chambers*	2B Frio	2A Karnes*	3A Nacogdoches*
4A Van Buren	2A Cherokee*	3B Gaines	3A Kaufman*	3A Navarro*
4A Warren	3B Childress	2A Galveston*	3A Kendall*	2A Newton*
4A Washington	3A Clay	3B Garza	2A Kenedy*	3B Nolan
4A Wayne	4B Cochran	3A Gillespie*	3B Kent	2A Nueces*
4A Weakley	3B Coke	3B Glasscock	3B Kerr	4B Ochiltree
4A White	3B Coleman	2A Goliad*	3B Kimble	4B Oldham
4A Williamson	3A Collin*	2A Gonzales*	3B King	2A Orange*
4A Wilson	3B Collingsworth	4B Gray	2B Kinney	3A Palo Pinto*
TEXAS	2A Colorado*	3A Grayson	2A Kleberg*	3A Panola*
2A Anderson*	2A Comal*	3A Gregg*	3B Knox	3A Parker*
3B Andrews	3A Comanche*	2A Grimes*	3A Lamar*	4B Parmer
2A Angelina*	3B Concho	2A Guadalupe*	4B Lamb	3B Pecos
2A Aransas*	3A Cooke	4B Hale	3A Lampasas*	2A Polk*
3A Archer	2A Coryell*	3B Hall	2B La Salle	4B Potter
4B Armstrong	3B Cottle	3A Hamilton*	2A Lavaca*	3B Presidio
2A Atascosa*	3B Crane	4B Hansford	2A Lee*	3A Rains*
2A Austin*	3B Crockett	3B Hardeman	2A Leon*	4B Randall
4B Bailey	3B Crosby	2A Hardin*	2A Liberty*	3B Reagan
2B Bandera	3B Culberson	2A Harris*	2A Limestone*	2B Real
2A Bastrop*	4B Dallam	3A Harrison*	4B Lipscomb	3A Red River*
3B Baylor	3A Dallas*	4B Hartley	2A Live Oak*	3B Reeves
2A Bee*	3B Dawson	3B Haskell	3A Llano*	2A Refugio*
2A Bell*	4B Deaf Smith	2A Hays*	3B Loving	4B Roberts
2A Bexar*	3A Delta	3B Hemphill	3B Lubbock	2A Robertson*
3A Blanco*	3A Denton*	3A Henderson*	3B Lynn	3A Rockwall*
3B Borden	2A DeWitt*	2A Hidalgo*	2A Madison*	3B Runnels
2A Bosque*	3B Dickens	2A Hill*	3A Marion*	3A Rusk*
3A Bowie*	2B Dimmit	4B Hockley	3B Martin	3A Sabine*
2A Brazoria*	4B Donley	3A Hood*	3B Mason	3A San Augustine*
2A Brazos*	2A Duval*	3A Hopkins*	2A Matagorda*	2A San Jacinto*
	3A Eastland	2A Houston*	2B Maverick	2A San Patricio*

(continued)

GENERAL REQUIREMENTS

TABLE R301.1—continued
CLIMATE ZONES, MOISTURE REGIMES, AND WARM-HUMID
DESIGNATIONS BY STATE, COUNTY AND TERRITORY

3A San Saba*	3A Young	4C Clark	4A Gilmer	**WISCONSIN**
3B Schleicher	2B Zapata	5B Columbia	5A Grant	6A Adams
3B Scurry	2B Zavala	4C Cowlitz	5A Greenbrier	7 Ashland
3B Shackelford	**UTAH**	5B Douglas	5A Hampshire	6A Barron
3A Shelby*	5B Beaver	6B Ferry	5A Hancock	7 Bayfield
4B Sherman	6B Box Elder	5B Franklin	5A Hardy	6A Brown
3A Smith*	6B Cache	5B Garfield	5A Harrison	6A Buffalo
3A Somervell*	6B Carbon	5B Grant	4A Jackson	7 Burnett
2A Starr*	6B Daggett	4C Grays Harbor	4A Jefferson	6A Calumet
3A Stephens	5B Davis	4C Island	4A Kanawha	6A Chippewa
3B Sterling	6B Duchesne	4C Jefferson	5A Lewis	6A Clark
3B Stonewall	5B Emery	4C King	4A Lincoln	6A Columbia
3B Sutton	5B Garfield	4C Kitsap	4A Logan	6A Crawford
4B Swisher	5B Grand	5B Kittitas	5A Marion	6A Dane
3A Tarrant*	5B Iron	5B Klickitat	5A Marshall	6A Dodge
3B Taylor	5B Juab	4C Lewis	4A Mason	6A Door
3B Terrell	5B Kane	5B Lincoln	4A McDowell	7 Douglas
3B Terry	5B Millard	4C Mason	4A Mercer	6A Dunn
3B Throckmorton	6B Morgan	6B Okanogan	5A Mineral	6A Eau Claire
3A Titus*	5B Piute	4C Pacific	4A Mingo	7 Florence
3B Tom Green	6B Rich	6B Pend Oreille	5A Monongalia	6A Fond du Lac
2A Travis*	5B Salt Lake	4C Pierce	4A Monroe	7 Forest
2A Trinity*	5B San Juan	4C San Juan	4A Morgan	6A Grant
2A Tyler*	5B Sanpete	4C Skagit	5A Nicholas	6A Green
3A Upshur*	5B Sevier	5B Skamania	5A Ohio	6A Green Lake
3B Upton	6B Summit	4C Snohomish	5A Pendleton	6A Iowa
2B Uvalde	5B Tooele	5B Spokane	4A Pleasants	7 Iron
2B Val Verde	6B Uintah	6B Stevens	5A Pocahontas	6A Jackson
3A Van Zandt*	5B Utah	4C Thurston	5A Preston	6A Jefferson
2A Victoria*	6B Wasatch	4C Wahkiakum	4A Putnam	6A Juneau
2A Walker*	3B Washington	5B Walla Walla	5A Raleigh	6A Kenosha
2A Waller*	5B Wayne	4C Whatcom	5A Randolph	6A Kewaunee
3B Ward	5B Weber	5B Whitman	4A Ritchie	6A La Crosse
2A Washington*	**VERMONT**	5B Yakima	4A Roane	6A Lafayette
2B Webb	6A (all)	**WEST VIRGINIA**	5A Summers	7 Langlade
2A Wharton*	**VIRGINIA**	5A Barbour	5A Taylor	7 Lincoln
3B Wheeler	4A (all)	4A Berkeley	5A Tucker	6A Manitowoc
3A Wichita	**WASHINGTON**	4A Boone	4A Tyler	6A Marathon
3B Wilbarger	5B Adams	4A Braxton	5A Upshur	6A Marinette
2A Willacy*	5B Asotin	5A Brooke	4A Wayne	6A Marquette
2A Williamson*	5B Benton	4A Cabell	5A Webster	6A Menominee
2A Wilson*	5B Chelan	4A Calhoun	5A Wetzel	6A Milwaukee
3B Winkler	4C Clallam	4A Clay	4A Wirt	6A Monroe
3A Wise		5A Doddridge	4A Wood	6A Oconto
3A Wood*		5A Fayette	4A Wyoming	7 Oneida
4B Yoakum				6A Outagamie

(continued)

TABLE R301.1—continued
CLIMATE ZONES, MOISTURE REGIMES, AND WARM-HUMID DESIGNATIONS BY STATE, COUNTY AND TERRITORY

6A Ozaukee	7 Taylor	6B Big Horn	6B Sheridan	**NORTHERN MARIANA ISLANDS**
6A Pepin	6A Trempealeau	6B Campbell	7 Sublette	
6A Pierce	6A Vernon	6B Carbon	6B Sweetwater	
6A Polk	7 Vilas	6B Converse	7 Teton	1A (all)*
6A Portage	6A Walworth	6B Crook	6B Uinta	**PUERTO RICO**
7 Price	7 Washburn	6B Fremont	6B Washakie	
6A Racine	6A Washington	5B Goshen	6B Weston	1A (all)*
6A Richland	6A Waukesha	6B Hot Springs	**US TERRITORIES**	**VIRGIN ISLANDS**
6A Rock	6A Waupaca	6B Johnson		1A (all)*
6A Rusk	6A Waushara	6B Laramie	**AMERICAN SAMOA**	
6A Sauk	6A Winnebago	7 Lincoln		
7 Sawyer	6A Wood	6B Natrona	1A (all)*	
6A Shawano		6B Niobrara	**GUAM**	
6A Sheboygan	**WYOMING**	6B Park		
6A St. Croix	6B Albany	5B Platte	1A (all)*	

TABLE R301.3(1)
INTERNATIONAL CLIMATE ZONE DEFINITIONS

MAJOR CLIMATE TYPE DEFINITIONS
Marine (C) Definition—Locations meeting all four criteria: 1. Mean temperature of coldest month between -3°C (27°F) and 18°C (65°F). 2. Warmest month mean < 22°C (72°F). 3. At least four months with mean temperatures over 10°C (50°F). 4. Dry season in summer. The month with the heaviest precipitation in the cold season has at least three times as much precipitation as the month with the least precipitation in the rest of the year. The cold season is October through March in the Northern Hemisphere and April through September in the Southern Hemisphere.
Dry (B) Definition—Locations meeting the following criteria: Not marine and $P_{in} < 0.44 \times (TF - 19.5)$ $[P_{cm} < 2.0 \times (TC + 7)$ in SI units] where: P_{in} = Annual precipitation in inches (cm) T = Annual mean temperature in °F (°C)
Moist (A) Definition—Locations that are not marine and not dry.
Warm-humid Definition—Moist (A) locations where either of the following wet-bulb temperature conditions shall occur during the warmest six consecutive months of the year: 1. 67°F (19.4°C) or higher for 3,000 or more hours; or 2. 73°F (22.8°C) or higher for 1,500 or more hours.

For SI: °C = [(°F)-32]/1.8, 1 inch = 2.54 cm.

GENERAL REQUIREMENTS

TABLE R301.3(2)
INTERNATIONAL CLIMATE ZONE DEFINITIONS

ZONE NUMBER	THERMAL CRITERIA	
	IP Units	SI Units
1	9000 < CDD50°F	5000 < CDD10°C
2	6300 < CDD50°F ≤ 9000	3500 < CDD10°C ≤ 5000
3A and 3B	4500 < CDD50°F ≤ 6300 AND HDD65°F ≤ 5400	2500 < CDD10°C ≤ 3500 AND HDD18°C ≤ 3000
4A and 4B	CDD50°F ≤ 4500 AND HDD65°F ≤ 5400	CDD10°C ≤ 2500 AND HDD18°C ≤ 3000
3C	HDD65°F ≤ 3600	HDD18°C ≤ 2000
4C	3600 < HDD65°F ≤ 5400	2000 < HDD18°C ≤ 3000
5	5400 < HDD65°F ≤ 7200	3000 < HDD18°C ≤ 4000
6	7200 < HDD65°F ≤ 9000	4000 < HDD18°C ≤ 5000
7	9000 < HDD65°F ≤ 12600	5000 < HDD18°C ≤ 7000
8	12600 < HDD65°F	7000 < HDD18°C

For SI: °C = [(°F)-32]/1.8.

SECTION R302
DESIGN CONDITIONS

R302.1 Interior design conditions. The interior design temperatures used for heating and cooling load calculations shall be a maximum of 72°F (22°C) for heating and minimum of 75°F (24°C) for cooling.

SECTION R303
MATERIALS, SYSTEMS AND EQUIPMENT

R303.1 Identification. Materials, systems and equipment shall be identified in a manner that will allow a determination of compliance with the applicable provisions of this code.

R303.1.1 Building thermal envelope insulation. An *R*-value identification mark shall be applied by the manufacturer to each piece of *building thermal envelope* insulation 12 inches (305 mm) or greater in width. Alternately, the insulation installers shall provide a certification listing the type, manufacturer and *R*-value of insulation installed in each element of the *building thermal envelope*. For blown or sprayed insulation (fiberglass and cellulose), the initial installed thickness, settled thickness, settled *R*-value, installed density, coverage area and number of bags installed shall be *listed* on the certification. For sprayed polyurethane foam (SPF) insulation, the installed thickness of the areas covered and *R*-value of installed thickness shall be *listed* on the certification. For insulated siding, the *R*-value shall be labeled on the product's package and shall be *listed* on the certification. The insulation installer shall sign, date and post the certification in a conspicuous location on the job site.

R303.1.1.1 Blown or sprayed roof/ceiling insulation. The thickness of blown-in or sprayed roof/ceiling insulation (fiberglass or cellulose) shall be written in inches (mm) on markers that are installed at least one for every 300 square feet (28 m^2) throughout the attic space. The markers shall be affixed to the trusses or joists and marked with the minimum initial installed thickness with numbers not less than 1 inch (25 mm) in height. Each marker shall face the attic access opening. Spray polyurethane foam thickness and installed *R*-value shall be *listed* on certification provided by the insulation installer.

R303.1.2 Insulation mark installation. Insulating materials shall be installed such that the manufacturer's *R*-value mark is readily observable upon inspection.

R303.1.3 Fenestration product rating. *U*-factors of fenestration products (windows, doors and skylights) shall be determined in accordance with NFRC 100.

> **Exception:** Where required, garage door *U*-factors shall be determined in accordance with either NFRC 100 or ANSI/DASMA 105.

U-factors shall be determined by an accredited, independent laboratory, and *labeled* and certified by the manufacturer.

Products lacking such a *labeled U*-factor shall be assigned a default *U*-factor from Table R303.1.3(1) or R303.1.3(2). The solar heat gain coefficient (SHGC) and *visible transmittance* (VT) of glazed fenestration products (windows, glazed doors and skylights) shall be determined

TABLE R303.1.3(1)
DEFAULT GLAZED FENESTRATION *U*-FACTORS

FRAME TYPE	SINGLE PANE	DOUBLE PANE	SKYLIGHT	
			Single	Double
Metal	1.20	0.80	2.00	1.30
Metal with Thermal Break	1.10	0.65	1.90	1.10
Nonmetal or Metal Clad	0.95	0.55	1.75	1.05
Glazed Block	0.60			

TABLE R303.1.3(2)
DEFAULT DOOR *U*-FACTORS

DOOR TYPE	*U*-FACTOR
Uninsulated Metal	1.20
Insulated Metal	0.60
Wood	0.50
Insulated, nonmetal edge, max 45% glazing, any glazing double pane	0.35

in accordance with NFRC 200 by an accredited, independent laboratory, and *labeled* and certified by the manufacturer. Products lacking such a *labeled* SHGC or VT shall be assigned a default SHGC or VT from Table R303.1.3(3).

TABLE R303.1.3(3)
DEFAULT GLAZED FENESTRATION SHGC AND VT

	SINGLE GLAZED		DOUBLE GLAZED		GLAZED BLOCK
	Clear	Tinted	Clear	Tinted	
SHGC	0.8	0.7	0.7	0.6	0.6
VT	0.6	0.3	0.6	0.3	0.6

R303.1.4 Insulation product rating. The thermal resistance (*R*-value) of insulation shall be determined in accordance with the U.S. Federal Trade Commission *R*-value rule (CFR Title 16, Part 460) in units of h · ft^2 · °F/Btu at a mean temperature of 75°F (24°C).

R303.1.4.1 Insulated siding. The thermal resistance (*R*-value) of insulated siding shall be determined in accordance with ASTM C 1363. Installation for testing shall be in accordance with the manufacturer's instructions.

R303.2 Installation. Materials, systems and equipment shall be installed in accordance with the manufacturer's instructions and the *International Building Code* or *International Residential Code*, as applicable.

R303.2.1 Protection of exposed foundation insulation. Insulation applied to the exterior of basement walls, crawlspace walls and the perimeter of slab-on-grade floors shall have a rigid, opaque and weather-resistant protective covering to prevent the degradation of the insulation's thermal performance. The protective covering shall cover the exposed exterior insulation and extend not less than 6 inches (153 mm) below grade.

R303.3 Maintenance information. Maintenance instructions shall be furnished for equipment and systems that require preventive maintenance. Required regular maintenance actions shall be clearly stated and incorporated on a readily accessible label. The label shall include the title or publication number for the operation and maintenance manual for that particular model and type of product.

CHAPTER 4 [RE]
RESIDENTIAL ENERGY EFFICIENCY

SECTION R401
GENERAL

R401.1 Scope. This chapter applies to residential buildings.

R401.2 Compliance. Projects shall comply with one of the following:

1. Sections R401 through R404.
2. Section R405 and the provisions of Sections R401 through R404 labeled "Mandatory."
3. An energy rating index (ERI) approach in Section R406.

R401.2.1 Tropical zone. *Residential buildings* in the tropical zone at elevations below 2,400 feet (731.5 m) above sea level shall be deemed to comply with this chapter where the following conditions are met:

1. Not more than one-half of the *occupied* space is air conditioned.
2. The *occupied* space is not heated.
3. Solar, wind or other renewable energy source supplies not less than 80 percent of the energy for service water heating.
4. Glazing in *conditioned* space has a *solar heat gain coefficient* of less than or equal to 0.40, or has an overhang with a projection factor equal to or greater than 0.30.
5. Permanently installed lighting is in accordance with Section R404.
6. The exterior roof surface complies with one of the options in Table C402.3 or the roof/ceiling has insulation with an *R-value* of R-15 or greater. If present, attics above the insulation are vented and attics below the insulation are unvented.
7. Roof surfaces have a minimum slope of $^1/_4$ inch per foot of run. The finished roof does not have water accumulation areas.
8. Operable fenestration provides ventilation area equal to not less than 14 percent of the floor area in each room. Alternatively, equivalent ventilation is provided by a ventilation fan.
9. Bedrooms with exterior walls facing two different directions have operable fenestration on exterior walls facing two directions.
10. Interior doors to bedrooms are capable of being secured in the open position.
11. A ceiling fan or ceiling fan rough-in is provided for bedrooms and the largest space that is not used as a bedroom.

R401.3 Certificate (Mandatory). A permanent certificate shall be completed by the builder or registered design professional and posted on a wall in the space where the furnace is located, a utility room or an approved location inside the building. Where located on an electrical panel, the certificate shall not cover or obstruct the visibility of the circuit directory label, service disconnect label or other required labels. The certificate shall list the predominant *R*-values of insulation installed in or on ceiling/roof, walls, foundation (slab, basement wall, crawlspace wall and floor) and ducts outside conditioned spaces; *U*-factors for fenestration and the solar heat gain coefficient (SHGC) of fenestration, and the results from any required duct system and building envelope air leakage testing done on the building. Where there is more than one value for each component, the certificate shall list the value covering the largest area. The certificate shall list the types and efficiencies of heating, cooling and service water heating equipment. Where a gas-fired unvented room heater, electric furnace or baseboard electric heater is installed in the residence, the certificate shall list "gas-fired unvented room heater," "electric furnace" or "baseboard electric heater," as appropriate. An efficiency shall not be *listed* for gas-fired unvented room heaters, electric furnaces or electric baseboard heaters.

SECTION R402
BUILDING THERMAL ENVELOPE

R402.1 General (Prescriptive). The *building thermal envelope* shall meet the requirements of Sections R402.1.1 through R402.1.5.

> **Exception:** The following low-energy buildings, or portions thereof, separated from the remainder of the building by *building thermal envelope* assemblies complying with this section shall be exempt from the *building thermal envelope* provisions of Section R402.
>
> 1. Those with a peak design rate of energy usage less than 3.4 Btu/h·ft^2 (10.7 W/m^2) or 1.0 watt/ft^2 of floor area for space-conditioning purposes.
> 2. Those that do not contain *conditioned space*.

R402.1.1 Vapor retarder. Wall assemblies in the *building thermal envelope* shall comply with the vapor retarder requirements of Section R702.7 of the *International Residential Code* or Section 1405.3 of the *International Building Code,* as applicable.

R402.1.2 Insulation and fenestration criteria. The *building thermal envelope* shall meet the requirements of Table R402.1.2, based on the climate zone specified in Chapter 3.

R402.1.3 *R*-value computation. Insulation material used in layers, such as framing cavity insulation, or continuous insulation shall be summed to compute the corresponding component *R*-value. The manufacturer's settled *R*-value shall be used for blown insulation. Computed *R*-values shall not include an *R*-value for other building materials or air films. Where insulated siding is used for the purpose of complying with the continuous insulation requirements of Table R402.1.2, the manufacturer's labeled *R*-value for insulated siding shall be reduced by R-0.6.

R402.1.4 *U*-factor alternative. An assembly with a *U*-factor equal to or less than that specified in Table R402.1.4 shall be permitted as an alternative to the *R*-value in Table R402.1.2.

R402.1.5 Total UA alternative. If the total *building thermal envelope* UA (sum of *U*-factor times assembly area) is less than or equal to the total UA resulting from using the *U*-factors in Table R402.1.4 (multiplied by the same assembly area as in the proposed building), the building shall be considered in compliance with Table R402.1.2. The UA calculation shall be done using a method consistent with the ASHRAE *Handbook of Fundamentals* and shall include the thermal bridging effects of framing materials. The SHGC requirements shall be met in addition to UA compliance.

R402.2 Specific insulation requirements (Prescriptive). In addition to the requirements of Section R402.1, insulation shall meet the specific requirements of Sections R402.2.1 through R402.2.13.

R402.2.1 Ceilings with attic spaces. Where Section R402.1.2 would require R-38 insulation in the ceiling,

TABLE R402.1.2
INSULATION AND FENESTRATION REQUIREMENTS BY COMPONENT[a]

CLIMATE ZONE	FENESTRATION *U*-FACTOR[b]	SKYLIGHT[b] *U*-FACTOR	GLAZED FENESTRATION SHGC[b, e]	CEILING *R*-VALUE	WOOD FRAME WALL *R*-VALUE	MASS WALL *R*-VALUE[i]	FLOOR *R*-VALUE	BASEMENT[c] WALL *R*-VALUE	SLAB[d] *R*-VALUE & DEPTH	CRAWL SPACE[c] WALL *R*-VALUE
1	NR	0.75	0.25	30	13	3/4	13	0	0	0
2	0.40	0.65	0.25	38	13	4/6	13	0	0	0
3	0.35	0.55	0.25	38	20 or 13+5[h]	8/13	19	5/13[f]	0	5/13
4 except Marine	0.35	0.55	0.40	49	20 or 13+5[h]	8/13	19	10/13	10, 2 ft	10/13
5 and Marine 4	0.32	0.55	NR	49	20 or 13+5[h]	13/17	30[g]	15/19	10, 2 ft	15/19
6	0.32	0.55	NR	49	20+5 or 13+10[h]	15/20	30[g]	15/19	10, 4 ft	15/19
7 and 8	0.32	0.55	NR	49	20+5 or 13+10[h]	19/21	38[g]	15/19	10, 4 ft	15/19

For SI: 1 foot = 304.8 mm.

a. R-values are minimums. *U*-factors and SHGC are maximums. When insulation is installed in a cavity which is less than the label or design thickness of the insulation, the installed *R*-value of the insulation shall not be less than the *R*-value specified in the table.
b. The fenestration *U*-factor column excludes skylights. The SHGC column applies to all glazed fenestration. Exception: Skylights may be excluded from glazed fenestration SHGC requirements in climate zones 1 through 3 where the SHGC for such skylights does not exceed 0.30.
c. "15/19" means R-15 continuous insulation on the interior or exterior of the home or R-19 cavity insulation at the interior of the basement wall. "15/19" shall be permitted to be met with R-13 cavity insulation on the interior of the basement wall plus R-5 continuous insulation on the interior or exterior of the home. "10/13" means R-10 continuous insulation on the interior or exterior of the home or R-13 cavity insulation at the interior of the basement wall.
d. R-5 shall be added to the required slab edge *R*-values for heated slabs. Insulation depth shall be the depth of the footing or 2 feet, whichever is less in Climate Zones 1 through 3 for heated slabs.
e. There are no SHGC requirements in the Marine Zone.
f. Basement wall insulation is not required in warm-humid locations as defined by Figure R301.1 and Table R301.1.
g. Or insulation sufficient to fill the framing cavity, R-19 minimum.
h. The first value is cavity insulation, the second value is continuous insulation, so "13+5" means R-13 cavity insulation plus R-5 continuous insulation.
i. The second *R*-value applies when more than half the insulation is on the interior of the mass wall.

TABLE R402.1.4
EQUIVALENT *U*-FACTORS[a]

CLIMATE ZONE	FENESTRATION *U*-FACTOR	SKYLIGHT *U*-FACTOR	CEILING *U*-FACTOR	FRAME WALL *U*-FACTOR	MASS WALL *U*-FACTOR[b]	FLOOR *U*-FACTOR	BASEMENT WALL *U*-FACTOR	CRAWL SPACE WALL *U*-FACTOR
1	0.50	0.75	0.035	0.084	0.197	0.064	0.360	0.477
2	0.40	0.65	0.030	0.084	0.165	0.064	0.360	0.477
3	0.35	0.55	0.030	0.060	0.098	0.047	0.091[c]	0.136
4 except Marine	0.35	0.55	0.026	0.060	0.098	0.047	0.059	0.065
5 and Marine 4	0.32	0.55	0.026	0.060	0.082	0.033	0.050	0.055
6	0.32	0.55	0.026	0.045	0.060	0.033	0.050	0.055
7 and 8	0.32	0.55	0.026	0.045	0.057	0.028	0.050	0.055

a. Nonfenestration *U*-factors shall be obtained from measurement, calculation or an approved source.
b. When more than half the insulation is on the interior, the mass wall U-factors shall be a maximum of 0.17 in Climate Zone 1, 0.14 in Climate Zone 2, 0.12 in Climate Zone 3, 0.087 in Climate Zone 4 except Marine, 0.065 in Climate Zone 5 and Marine 4, and 0.057 in Climate Zones 6 through 8.
c. Basement wall *U*-factor of 0.360 in warm-humid locations as defined by Figure R301.1 and Table R301.1.

installing R-30 over 100 percent of the ceiling area requiring insulation shall be deemed to satisfy the requirement for R-38 wherever the full height of uncompressed R-30 insulation extends over the wall top plate at the eaves. Similarly, where Section R402.1.2 would require R-49 insulation in the ceiling, installing R-38 over 100 percent of the ceiling area requiring insulation shall be deemed to satisfy the requirement for R-49 insulation wherever the full height of uncompressed R-38 insulation extends over the wall top plate at the eaves. This reduction shall not apply to the *U*-factor alternative approach in Section R402.1.4 and the total UA alternative in Section R402.1.5.

R402.2.2 Ceilings without attic spaces. Where Section R402.1.2 would require insulation levels above R-30 and the design of the roof/ceiling assembly does not allow sufficient space for the required insulation, the minimum required insulation for such roof/ceiling assemblies shall be R-30. This reduction of insulation from the requirements of Section R402.1.2 shall be limited to 500 square feet (46 m^2) or 20 percent of the total insulated ceiling area, whichever is less. This reduction shall not apply to the *U*-factor alternative approach in Section R402.1.4 and the total UA alternative in Section R402.1.5.

R402.2.3 Eave baffle. For air-permeable insulations in vented attics, a baffle shall be installed adjacent to soffit and eave vents. Baffles shall maintain an opening equal or greater than the size of the vent. The baffle shall extend over the top of the attic insulation. The baffle shall be permitted to be any solid material.

R402.2.4 Access hatches and doors. Access doors from conditioned spaces to unconditioned spaces such as attics and crawl spaces shall be weatherstripped and insulated to a level equivalent to the insulation on the surrounding surfaces. Access shall be provided to all equipment that prevents damaging or compressing the insulation. A wood-framed or equivalent baffle or retainer is required to be provided when loose-fill insulation is installed, the purpose of which is to prevent the loose-fill insulation from spilling into the living space when the attic access is opened, and to provide a permanent means of maintaining the installed *R*-value of the loose-fill insulation.

> **Exception:** Vertical doors that provide access from conditioned to unconditioned spaces shall be permitted to meet the fenestration requirements of Table R402.1.2 based on the applicable climate zone specified in Chapter 3.

R402.2.5 Mass walls. Mass walls for the purposes of this chapter shall be considered above-grade walls of concrete block, concrete, insulated concrete form (ICF), masonry cavity, brick (other than brick veneer), earth (adobe, compressed earth block, rammed earth) and solid timber/logs, or any other walls having a heat capacity greater than or equal to 6 Btu/ft^2 × °F (123 kJ/m^2 × K).

R402.2.6 Steel-frame ceilings, walls and floors. Steel-frame ceilings, walls, and floors shall meet the insulation requirements of Table R402.2.6 or shall meet the *U*-factor requirements of Table R402.1.4. The calculation of the *U*-factor for a steel-frame envelope assembly shall use a series-parallel path calculation method.

R402.2.7 Walls with partial structural sheathing. Where Section R402.1.2 would require continuous insulation on exterior walls and structural sheathing covers 40 percent or less of the gross area of all exterior walls, the continuous insulation *R*-value shall be permitted to be reduced by an amount necessary to result in a consistent total sheathing thickness, but not more than R-3, on areas of the walls covered by structural sheathing. This reduction shall not apply to the *U*-factor alternative approach in Section R402.1.4 and the total UA alternative in Section R402.1.5.

TABLE R402.2.6
STEEL-FRAME CEILING, WALL AND FLOOR INSULATION (*R*-VALUE)

WOOD FRAME *R*-VALUE REQUIREMENT	COLD-FORMED STEEL EQUIVALENT *R*-VALUE[a]
Steel Truss Ceilings[b]	
R-30	R-38 or R-30 + 3 or R-26 + 5
R-38	R-49 or R-38 + 3
R-49	R-38 + 5
Steel Joist Ceilings[b]	
R-30	R-38 in 2 × 4 or 2 × 6 or 2 × 8 R-49 in any framing
R-38	R-49 in 2 × 4 or 2 × 6 or 2 × 8 or 2 × 10
Steel-Framed Wall, 16″ on center	
R-13	R-13 + 4.2 or R-19 + 2.1 or R-21 + 2.8 or R-0 + 9.3 or R-15 + 3.8 or R-21 + 3.1
R-13 + 3	R-0 + 11.2 or R-13 + 6.1 or R-15 + 5.7 or R-19 + 5.0 or R-21 + 4.7
R-20	R-0 + 14.0 or R-13 + 8.9 or R-15 + 8.5 or R-19 + 7.8 or R-19 + 6.2 or R-21 + 7.5
R-20 + 5	R-13 + 12.7 or R-15 + 12.3 or R-19 + 11.6 or R-21 + 11.3 or R-25 + 10.9
R-21	R-0 + 14.6 or R-13 + 9.5 or R-15 + 9.1 or R-19 + 8.4 or R-21 + 8.1 or R-25 + 7.7
Steel Framed Wall, 24″ on center	
R-13	R-0 + 9.3 or R-13 + 3.0 or R-15 + 2.4
R-13 + 3	R-0 + 11.2 or R-13 + 4.9 or R-15 + 4.3 or R-19 + 3.5 or R-21 + 3.1
R-20	R-0 + 14.0 or R-13 + 7.7 or R-15 + 7.1 or R-19 + 6.3 or R-21 + 5.9
R-20 + 5	R-13 + 11.5 or R-15 + 10.9 or R-19 + 10.1 or R-21 + 9.7 or R-25 + 9.1
R-21	R-0 + 14.6 or R-13 + 8.3 or R-15 + 7.7 or R-19 + 6.9 or R-21 + 6.5 or R-25 + 5.9
Steel Joist Floor	
R-13	R-19 in 2 × 6, or R-19 + 6 in 2 × 8 or 2 × 10
R-19	R-19 + 6 in 2 × 6, or R-19 + 12 in 2 × 8 or 2 × 10

a. Cavity insulation *R*-value is listed first, followed by continuous insulation *R*-value.

b. Insulation exceeding the height of the framing shall cover the framing.

R402.2.8 Floors. Floor framing-cavity insulation shall be installed to maintain permanent contact with the underside of the subfloor decking.

Exception: The floor framing-cavity insulation shall be permitted to be in contact with the topside of sheathing or continuous insulation installed on the bottom side of floor framing where combined with insulation that meets or exceeds the minimum wood frame wall *R*-value in Table 402.1.2 and that extends from the bottom to the top of all perimeter floor framing members.

R402.2.9 Basement walls. Walls associated with conditioned basements shall be insulated from the top of the *basement wall* down to 10 feet (3048 mm) below grade or to the basement floor, whichever is less. Walls associated with unconditioned basements shall meet this requirement unless the floor overhead is insulated in accordance with Sections R402.1.2 and R402.2.8.

R402.2.10 Slab-on-grade floors. Slab-on-grade floors with a floor surface less than 12 inches (305 mm) below grade shall be insulated in accordance with Table R402.1.2. The insulation shall extend downward from the top of the slab on the outside or inside of the foundation wall. Insulation located below grade shall be extended the distance provided in Table R402.1.2 by any combination of vertical insulation, insulation extending under the slab or insulation extending out from the building. Insulation extending away from the building shall be protected by pavement or by not less than 10 inches (254 mm) of soil. The top edge of the insulation installed between the *exterior wall* and the edge of the interior slab shall be permitted to be cut at a 45-degree (0.79 rad) angle away from the *exterior wall*. Slab-edge insulation is not required in jurisdictions designated by the *code official* as having a very heavy termite infestation.

R402.2.11 Crawl space walls. As an alternative to insulating floors over crawl spaces, crawl space walls shall be permitted to be insulated when the crawl space is not vented to the outside. Crawl space wall insulation shall be permanently fastened to the wall and extend downward from the floor to the finished grade level and then vertically and/or horizontally for at least an additional 24 inches (610 mm). Exposed earth in unvented crawl space foundations shall be covered with a continuous Class I vapor retarder in accordance with the *International Building Code* or *International Residential Code*, as applicable. All joints of the vapor retarder shall overlap by 6 inches (153 mm) and be sealed or taped. The edges of the vapor retarder shall extend not less than 6 inches (153 mm) up the stem wall and shall be attached to the stem wall.

R402.2.12 Masonry veneer. Insulation shall not be required on the horizontal portion of the foundation that supports a masonry veneer.

R402.2.13 Sunroom insulation. *Sunrooms* enclosing conditioned space shall meet the insulation requirements of this code.

Exception: For *sunrooms* with *thermal isolation*, and enclosing conditioned space, the following exceptions to the insulation requirements of this code shall apply:

1. The minimum ceiling insulation *R*-values shall be R-19 in *Climate Zones* 1 through 4 and R-24 in *Climate Zones* 5 through 8.
2. The minimum wall *R*-value shall be R-13 in all *climate zones*. Walls separating a *sunroom* with a *thermal isolation* from *conditioned space* shall meet the *building thermal envelope* requirements of this code.

R402.3 Fenestration (Prescriptive). In addition to the requirements of Section R402, fenestration shall comply with Sections R402.3.1 through R402.3.5.

R402.3.1 *U*-factor. An area-weighted average of fenestration products shall be permitted to satisfy the *U*-factor requirements.

R402.3.2 Glazed fenestration SHGC. An area-weighted average of fenestration products more than 50-percent glazed shall be permitted to satisfy the SHGC requirements.

Dynamic glazing shall be permitted to satisfy the SHGC requirements of Table R402.1.2 provided the ratio of the higher to lower labeled SHGC is greater than or equal to 2.4, and the *dynamic glazing* is automatically controlled to modulate the amount of solar gain into the space in multiple steps. *Dynamic glazing* shall be considered separately from other fenestration, and area-weighted averaging with other fenestration that is not dynamic glazing shall not be permitted.

Exception: *Dynamic glazing* is not required to comply with this section when both the lower and higher labeled SHGC already comply with the requirements of Table R402.1.2.

R402.3.3 Glazed fenestration exemption. Up to 15 square feet (1.4 m^2) of glazed fenestration per dwelling unit shall be permitted to be exempt from *U*-factor and SHGC requirements in Section R402.1.2. This exemption shall not apply to the *U*-factor alternative approach in Section R402.1.4 and the Total UA alternative in Section R402.1.5.

R402.3.4 Opaque door exemption. One side-hinged opaque door assembly up to 24 square feet (2.22 m^2) in area is exempted from the *U*-factor requirement in Section R402.1.4. This exemption shall not apply to the *U*-factor alternative approach in Section R402.1.4 and the total UA alternative in Section R402.1.5.

R402.3.5 Sunroom fenestration. *Sunrooms* enclosing *conditioned space* shall meet the fenestration requirements of this code.

Exception: For *sunrooms* with *thermal isolation* and enclosing *conditioned space* in Climate Zones 2 through 8, the maximum fenestration *U*-factor shall be 0.45 and the maximum skylight *U*-factor shall be 0.70.

New fenestration separating the *sunroom* with *thermal isolation* from *conditioned space* shall meet the *building thermal envelope* requirements of this code.

R402.4 Air leakage (Mandatory). The *building thermal envelope* shall be constructed to limit air leakage in accordance with the requirements of Sections R402.4.1 through R402.4.5.

R402.4.1 Building thermal envelope. The *building thermal envelope* shall comply with Sections R402.4.1.1 and R402.4.1.2. The sealing methods between dissimilar materials shall allow for differential expansion and contraction.

R402.4.1.1 Installation. The components of the *building thermal envelope* as listed in Table R402.4.1.1 shall be installed in accordance with the manufacturer's instructions and the criteria listed in Table R402.4.1.1, as applicable to the method of construction. Where required by the *code official*, an *approved* third party shall inspect all components and verify compliance.

R402.4.1.2 Testing. The building or dwelling unit shall be tested and verified as having an air leakage rate not exceeding five air changes per hour in Climate Zones 1 and 2, and three air changes per hour in Climate Zones 3 through 8. Testing shall be conducted in accordance with ASTM E 779 or ASTM E 1827 and reported at a pressure of 0.2 inch w.g. (50 Pascals). Where required by the *code official*, testing shall be conducted by an *approved* third party. A written report of the results of the test shall be signed by the party conducting the test and provided to the *code official*. Testing shall be performed at any time after creation of all penetrations of the *building thermal envelope*.

During testing:

1. Exterior windows and doors, fireplace and stove doors shall be closed, but not sealed, beyond the intended weatherstripping or other infiltration control measures.
2. Dampers including exhaust, intake, makeup air, backdraft and flue dampers shall be closed, but not sealed beyond intended infiltration control measures.
3. Interior doors, if installed at the time of the test, shall be open.
4. Exterior doors for continuous ventilation systems and heat recovery ventilators shall be closed and sealed.
5. Heating and cooling systems, if installed at the time of the test, shall be turned off.
6. Supply and return registers, if installed at the time of the test, shall be fully open.

R402.4.2 Fireplaces. New wood-burning fireplaces shall have tight-fitting flue dampers or doors, and outdoor combustion air. Where using tight-fitting doors on factory-built fireplaces listed and labeled in accordance with UL 127, the doors shall be tested and listed for the fireplace. Where using tight-fitting doors on masonry fireplaces, the doors shall be listed and labeled in accordance with UL 907.

R402.4.3 Fenestration air leakage. Windows, skylights and sliding glass doors shall have an air infiltration rate of no more than 0.3 cfm per square foot (1.5 L/s/m^2), and swinging doors no more than 0.5 cfm per square foot (2.6 L/s/m^2), when tested according to NFRC 400 or AAMA/WDMA/CSA 101/I.S.2/A440 by an accredited, independent laboratory and *listed* and *labeled* by the manufacturer.

Exception: Site-built windows, skylights and doors.

R402.4.4 Rooms containing fuel-burning appliances. In Climate Zones 3 through 8, where open combustion air ducts provide combustion air to open combustion fuel burning appliances, the appliances and combustion air opening shall be located outside the building thermal envelope or enclosed in a room, isolated from inside the thermal envelope. Such rooms shall be sealed and insulated in accordance with the envelope requirements of Table R402.1.2, where the walls, floors and ceilings shall meet not less than the basement wall *R*-value requirement. The door into the room shall be fully gasketed and any water lines and ducts in the room insulated in accordance with Section R403. The combustion air duct shall be insulated where it passes through conditioned space to a minimum of R-8.

Exceptions:

1. Direct vent appliances with both intake and exhaust pipes installed continuous to the outside.
2. Fireplaces and stoves complying with Section R402.4.2 and Section R1006 of the *International Residential Code*.

R402.4.5 Recessed lighting. Recessed luminaires installed in the *building thermal envelope* shall be sealed to limit air leakage between conditioned and unconditioned spaces. All recessed luminaires shall be IC-rated and *labeled* as having an air leakage rate not more than 2.0 cfm (0.944 L/s) when tested in accordance with ASTM E 283 at a 1.57 psf (75 Pa) pressure differential. All recessed luminaires shall be sealed with a gasket or caulk between the housing and the interior wall or ceiling covering.

R402.5 Maximum fenestration *U*-factor and SHGC (Mandatory). The area-weighted average maximum fenestration *U*-factor permitted using tradeoffs from Section R402.1.5 or R405 shall be 0.48 in Climate Zones 4 and 5 and 0.40 in Climate Zones 6 through 8 for vertical fenestration, and 0.75 in Climate Zones 4 through 8 for skylights. The area-weighted average maximum fenestration SHGC permitted using tradeoffs from Section R405 in Climate Zones 1 through 3 shall be 0.50.

TABLE R402.4.1.1
AIR BARRIER AND INSULATION INSTALLATION

COMPONENT	AIR BARRIER CRITERIA	INSULATION INSTALLATION CRITERIA
General requirements	A continuous air barrier shall be installed in the building envelope. The exterior thermal envelope contains a continuous air barrier. Breaks or joints in the air barrier shall be sealed.	Air-permeable insulation shall not be used as a sealing material.
Ceiling/attic	The air barrier in any dropped ceiling/soffit shall be aligned with the insulation and any gaps in the air barrier shall be sealed. Access openings, drop down stairs or knee wall doors to unconditioned attic spaces shall be sealed.	The insulation in any dropped ceiling/soffit shall be aligned with the air barrier.
Walls	The junction of the foundation and sill plate shall be sealed. The junction of the top plate and the top of exterior walls shall be sealed. Knee walls shall be sealed.	Cavities within corners and headers of frame walls shall be insulated by completely filling the cavity with a material having a thermal resistance of R-3 per inch minimum. Exterior thermal envelope insulation for framed walls shall be installed in substantial contact and continuous alignment with the air barrier.
Windows, skylights and doors	The space between window/door jambs and framing, and skylights and framing shall be sealed.	
Rim joists	Rim joists shall include the air barrier.	Rim joists shall be insulated.
Floors (including above garage and cantilevered floors)	The air barrier shall be installed at any exposed edge of insulation.	Floor framing cavity insulation shall be installed to maintain permanent contact with the underside of subfloor decking, or floor framing cavity insulation shall be permitted to be in contact with the top side of sheathing, or continuous insulation installed on the underside of floor framing and extends from the bottom to the top of all perimeter floor framing members.
Crawl space walls	Exposed earth in unvented crawl spaces shall be covered with a Class I vapor retarder with overlapping joints taped.	Where provided instead of floor insulation, insulation shall be permanently attached to the crawlspace walls.
Shafts, penetrations	Duct shafts, utility penetrations, and flue shafts opening to exterior or unconditioned space shall be sealed.	
Narrow cavities		Batts in narrow cavities shall be cut to fit, or narrow cavities shall be filled by insulation that on installation readily conforms to the available cavity space.
Garage separation	Air sealing shall be provided between the garage and conditioned spaces.	
Recessed lighting	Recessed light fixtures installed in the building thermal envelope shall be sealed to the drywall.	Recessed light fixtures installed in the building thermal envelope shall be air tight and IC rated.
Plumbing and wiring		Batt insulation shall be cut neatly to fit around wiring and plumbing in exterior walls, or insulation that on installation readily conforms to available space shall extend behind piping and wiring.
Shower/tub on exterior wall	The air barrier installed at exterior walls adjacent to showers and tubs shall separate them from the showers and tubs.	Exterior walls adjacent to showers and tubs shall be insulated.
Electrical/phone box on exterior walls	The air barrier shall be installed behind electrical or communication boxes or air-sealed boxes shall be installed.	
HVAC register boots	HVAC register boots that penetrate building thermal envelope shall be sealed to the subfloor or drywall.	
Concealed sprinklers	When required to be sealed, concealed fire sprinklers shall only be sealed in a manner that is recommended by the manufacturer. Caulking or other adhesive sealants shall not be used to fill voids between fire sprinkler cover plates and walls or ceilings.	

a. In addition, inspection of log walls shall be in accordance with the provisions of ICC-400.

SECTION R403
SYSTEMS

R403.1 Controls (Mandatory). At least one thermostat shall be provided for each separate heating and cooling system.

R403.1.1 Programmable thermostat. The thermostat controlling the primary heating or cooling system of the dwelling unit shall be capable of controlling the heating and cooling system on a daily schedule to maintain different temperature set points at different times of the day. This thermostat shall include the capability to set back or temporarily operate the system to maintain *zone* temperatures down to 55°F (13°C) or up to 85°F (29°C). The thermostat shall initially be programmed by the manufacturer with a heating temperature set point no higher than 70°F (21°C) and a cooling temperature set point no lower than 78°F (26°C).

R403.1.2 Heat pump supplementary heat (Mandatory). Heat pumps having supplementary electric-resistance heat shall have controls that, except during defrost, prevent supplemental heat operation when the heat pump compressor can meet the heating load.

R403.2 Hot water boiler outdoor temperature setback. Hot water boilers that supply heat to the building through one- or two-pipe heating systems shall have an outdoor setback control that lowers the boiler water temperature based on the outdoor temperature.

R403.3 Ducts. Ducts and air handlers shall be in accordance with Sections R403.3.1 through R403.3.5.

R403.3.1 Insulation (Prescriptive). Supply and return ducts in attics shall be insulated to a minimum of R-8 where 3 inches (76 mm) in diameter and greater and R-6 where less than 3 inches (76 mm) in diameter. Supply and return ducts in other portions of the building shall be insulated to a minimum of R-6 where 3 inches (76 mm) in diameter or greater and R-4.2 where less than 3 inches (76 mm) in diameter.

Exception: Ducts or portions thereof located completely inside the *building thermal envelope.*

R403.3.2 Sealing (Mandatory). Ducts, air handlers and filter boxes shall be sealed. Joints and seams shall comply with either the *International Mechanical Code* or *International Residential Code*, as applicable.

Exceptions:
1. Air-impermeable spray foam products shall be permitted to be applied without additional joint seals.
2. For ducts having a static pressure classification of less than 2 inches of water column (500 Pa), additional closure systems shall not be required for continuously welded joints and seams, and locking-type joints and seams of other than the snap-lock and button-lock types.

R403.3.2.1 Sealed air handler. Air handlers shall have a manufacturer's designation for an air leakage of no more than 2 percent of the design air flow rate when tested in accordance with ASHRAE 193.

R403.3.3 Duct testing (Mandatory). Ducts shall be pressure tested to determine air leakage by one of the following methods:

1. Rough-in test: Total leakage shall be measured with a pressure differential of 0.1 inch w.g. (25 Pa) across the system, including the manufacturer's air handler enclosure if installed at the time of the test. All registers shall be taped or otherwise sealed during the test.

2. Postconstruction test: Total leakage shall be measured with a pressure differential of 0.1 inch w.g. (25 Pa) across the entire system, including the manufacturer's air handler enclosure. Registers shall be taped or otherwise sealed during the test.

Exception: A duct air leakage test shall not be required where the ducts and air handlers are located entirely within the building thermal envelope.

A written report of the results of the test shall be signed by the party conducting the test and provided to the *code official*.

R403.3.4 Duct leakage (Prescriptive). The total leakage of the ducts, where measured in accordance with Section R403.3.3, shall be as follows:

1. Rough-in test: The total leakage shall be less than or equal to 4 cubic feet per minute (113.3 L/min) per 100 square feet (9.29 m^2) of conditioned floor area where the air handler is installed at the time of the test. Where the air handler is not installed at the time of the test, the total leakage shall be less than or equal to 3 cubic feet per minute (85 L/min) per 100 square feet (9.29 m^2) of conditioned floor area.

2. Postconstruction test: Total leakage shall be less than or equal to 4 cubic feet per minute (113.3 L/min) per 100 square feet (9.29 m^2) of conditioned floor area.

R403.3.5 Building cavities (Mandatory). Building framing cavities shall not be used as ducts or plenums.

R403.4 Mechanical system piping insulation (Mandatory). Mechanical system piping capable of carrying fluids above 105°F (41°C) or below 55°F (13°C) shall be insulated to a minimum of R-3.

R403.4.1 Protection of piping insulation. Piping insulation exposed to weather shall be protected from damage, including that caused by sunlight, moisture, equipment maintenance and wind, and shall provide shielding from solar radiation that can cause degradation of the material. Adhesive tape shall not be permitted.

R403.5 Service hot water systems. Energy conservation measures for service hot water systems shall be in accordance with Sections R403.5.1 through R403.5.4.

R403.5.1 Heated water circulation and temperature maintenance systems (Mandatory). Heated water circulation systems shall be in accordance with Section R403.5.1.1. Heat trace temperature maintenance systems shall be in accordance with Section R403.5.1.2. Automatic controls, temperature sensors and pumps shall be accessible. Manual controls shall be readily accessible.

R403.5.1.1 Circulation systems. Heated water circulation systems shall be provided with a circulation pump. The system return pipe shall be a dedicated return pipe or a cold water supply pipe. Gravity and thermosyphon circulation systems shall be prohibited. Controls for circulating hot water system pumps shall start the pump based on the identification of a demand for hot water within the occupancy. The controls shall automatically turn off the pump when the water in the circulation loop is at the desired temperature and when there is no demand for hot water.

R403.5.1.2 Heat trace systems. Electric heat trace systems shall comply with IEEE 515.1 or UL 515. Controls for such systems shall automatically adjust the energy input to the heat tracing to maintain the desired water temperature in the piping in accordance with the times when heated water is used in the occupancy.

R403.5.2 Demand recirculation systems. A water distribution system having one or more recirculation pumps that pump water from a heated water supply pipe back to the heated water source through a cold water supply pipe shall be a *demand recirculation water system*. Pumps shall have controls that comply with both of the following:

1. The control shall start the pump upon receiving a signal from the action of a user of a fixture or appliance, sensing the presence of a user of a fixture or sensing the flow of hot or tempered water to a fixture fitting or appliance.

2. The control shall limit the temperature of the water entering the cold water piping to 104°F (40°C).

R403.5.3 Hot water pipe insulation (Prescriptive). Insulation for hot water pipe with a minimum thermal resistance (*R*-value) of R-3 shall be applied to the following:

1. Piping $^3/_4$ inch (19.1 mm) and larger in nominal diameter.
2. Piping serving more than one dwelling unit.
3. Piping located outside the conditioned space.
4. Piping from the water heater to a distribution manifold.
5. Piping located under a floor slab.
6. Buried in piping.
7. Supply and return piping in recirculation systems other than demand recirculation systems.

R403.5.4 Drain water heat recovery units. Drain water heat recovery units shall comply with CSA B55.2. Drain water heat recovery units shall be tested in accordance with CSA B55.1. Potable water-side pressure loss of drain water heat recovery units shall be less than 3 psi (20.7 kPa) for individual units connected to one or two showers. Potable water-side pressure loss of drain water heat recovery units shall be less than 2 psi (13.8 kPa) for individual units connected to three or more showers.

R403.6 Mechanical ventilation (Mandatory). The building shall be provided with ventilation that meets the requirements of the *International Residential Code* or *International Mechanical Code*, as applicable, or with other approved means of ventilation. Outdoor air intakes and exhausts shall have automatic or gravity dampers that close when the ventilation system is not operating.

R403.6.1 Whole-house mechanical ventilation system fan efficacy. Mechanical ventilation system fans shall meet the efficacy requirements of Table R403.6.1.

Exception: Where mechanical ventilation fans are integral to tested and listed HVAC equipment, they shall be powered by an electronically commutated motor.

R403.7 Equipment sizing and efficiency rating (Mandatory). Heating and cooling equipment shall be sized in accordance with ACCA Manual S based on building loads calculated in accordance with ACCA Manual J or other *approved* heating and cooling calculation methodologies. New or replacement heating and cooling equipment shall have an efficiency rating equal to or greater than the minimum required by federal law for the geographic location where the equipment is installed.

R403.8 Systems serving multiple dwelling units (Mandatory). Systems serving multiple dwelling units shall comply with Sections C403 and C404 of the IECC—Commercial Provisions in lieu of Section R403.

R403.9 Snow melt and ice system controls (Mandatory). Snow- and ice-melting systems, supplied through energy service to the building, shall include automatic controls capable of shutting off the system when the pavement temperature is above 50°F (10°C), and no precipitation is falling and an automatic or manual control that will allow shutoff when the outdoor temperature is above 40°F (4.8°C).

TABLE R403.6.1
MECHANICAL VENTILATION SYSTEM FAN EFFICACY

FAN LOCATION	AIR FLOW RATE MINIMUM (CFM)	MINIMUM EFFICACY (CFM/WATT)	AIR FLOW RATE MAXIMUM (CFM)
Range hoods	Any	2.8 cfm/watt	Any
In-line fan	Any	2.8 cfm/watt	Any
Bathroom, utility room	10	1.4 cfm/watt	< 90
Bathroom, utility room	90	2.8 cfm/watt	Any

For SI: 1 cfm = 28.3 L/min.

R403.10 Pools and permanent spa energy consumption (Mandatory). The energy consumption of pools and permanent spas shall be in accordance with Sections R403.10.1 through R403.10.3.

R403.10.1 Heaters. The electric power to heaters shall be controlled by a readily *accessible* on-off switch that is an integral part of the heater mounted on the exterior of the heater, or external to and within 3 feet (914 mm) of the heater. Operation of such switch shall not change the setting of the heater thermostat. Such switches shall be in addition to a circuit breaker for the power to the heater. Gas-fired heaters shall not be equipped with continuously burning ignition pilots.

R403.10.2 Time switches. Time switches or other control methods that can automatically turn off and on according to a preset schedule shall be installed for heaters and pump motors. Heaters and pump motors that have built-in time switches shall be in compliance with this section.

Exceptions:

1. Where public health standards require 24-hour pump operation.
2. Pumps that operate solar- and waste-heat-recovery pool heating systems.

R403.10.3 Covers. Outdoor heated pools and outdoor permanent spas shall be provided with a vapor-retardant cover or other *approved* vapor-retardant means.

Exception: Where more than 70 percent of the energy for heating, computed over an operation season, is from site-recovered energy, such as from a heat pump or solar energy source, covers or other vapor-retardant means shall not be required.

R403.11 Portable spas (Mandatory). The energy consumption of electric-powered portable spas shall be controlled by the requirements of APSP-14.

R403.12 Residential pools and permanent residential spas. Residential swimming pools and permanent residential spas that are accessory to detached one- and two-family dwellings and townhouses three stories or less in height above grade plane and that are available only to the household and its guests shall be in accordance with APSP-15.

SECTION R404
ELECTRICAL POWER AND LIGHTING SYSTEMS

R404.1 Lighting equipment (Mandatory). Not less than 75 percent of the lamps in permanently installed lighting fixtures shall be high-efficacy lamps or not less than 75 percent of the permanently installed lighting fixtures shall contain only high-efficacy lamps.

Exception: Low-voltage lighting.

R404.1.1 Lighting equipment (Mandatory). Fuel gas lighting systems shall not have continuously burning pilot lights.

SECTION R405
SIMULATED PERFORMANCE ALTERNATIVE
(PERFORMANCE)

R405.1 Scope. This section establishes criteria for compliance using simulated energy performance analysis. Such analysis shall include heating, cooling and service water heating energy only.

R405.2 Mandatory requirements. Compliance with this section requires that the mandatory provisions identified in Section R401.2 be met. All supply and return ducts not completely inside the *building thermal envelope* shall be insulated to a minimum of R-6.

R405.3 Performance-based compliance. Compliance based on simulated energy performance requires that a proposed residence (*proposed design*) be shown to have an annual energy cost that is less than or equal to the annual energy cost of the *standard reference design*. Energy prices shall be taken from a source *approved* by the *code official,* such as the Department of Energy, Energy Information Administration's *State Energy Price and Expenditure Report. Code officials* shall be permitted to require time-of-use pricing in energy cost calculations.

Exception: The energy use based on source energy expressed in Btu or Btu per square foot of *conditioned floor area* shall be permitted to be substituted for the energy cost. The source energy multiplier for electricity shall be 3.16. The source energy multiplier for fuels other than electricity shall be 1.1.

R405.4 Documentation. Documentation of the software used for the performance design and the parameters for the building shall be in accordance with Sections R405.4.1 through R405.4.3.

R405.4.1 Compliance software tools. Documentation verifying that the methods and accuracy of the compliance software tools conform to the provisions of this section shall be provided to the *code official.*

R405.4.2 Compliance report. Compliance software tools shall generate a report that documents that the *proposed design* complies with Section R405.3. A compliance report on the *proposed design* shall be submitted with the application for the building permit. Upon completion of the building, a compliance report based on the as-built condition of the building shall be submitted to the *code official* before a certificate of occupancy is issued. Batch sampling of buildings to determine energy code compliance for all buildings in the batch shall be prohibited.

Compliance reports shall include information in accordance with Sections R405.4.2.1 and R405.4.2.2. Where the *proposed design* of a building could be built on different sites where the cardinal orientation of the building on each site is different, compliance of the *proposed design* for the purposes of the application for the building permit shall be based on the worst-case orientation, worst-case configuration, worst-case building air leakage and worst-case duct leakage. Such worst-case parameters shall be used as inputs to the compliance software for energy analysis.

R405.4.2.1 Compliance report for permit application. A compliance report submitted with the application for building permit shall include the following:

1. Building street address, or other building site identification.
2. A statement indicating that the *proposed design* complies with Section R405.3.
3. An inspection checklist documenting the building component characteristics of the *proposed design* as indicated in Table R405.5.2(1). The inspection checklist shall show results for both the *standard reference design* and the *proposed design* with user inputs to the compliance software to generate the results.
4. A site-specific energy analysis report that is in compliance with Section R405.3.
5. The name of the individual performing the analysis and generating the report.
6. The name and version of the compliance software tool.

R405.4.2.2 Compliance report for certificate of occupancy. A compliance report submitted for obtaining the certificate of occupancy shall include the following:

1. Building street address, or other building site identification.
2. A statement indicating that the as-built building complies with Section R405.3.
3. A certificate indicating that the building passes the performance matrix for code compliance and listing the energy saving features of the buildings.
4. A site-specific energy analysis report that is in compliance with Section R405.3.
5. The name of the individual performing the analysis and generating the report.
6. The name and version of the compliance software tool.

R405.4.3 Additional documentation. The *code official* shall be permitted to require the following documents:

1. Documentation of the building component characteristics of the *standard reference design*.
2. A certification signed by the builder providing the building component charactcristics of the *proposed design* as given in Table R405.5.2(1).
3. Documentation of the actual values used in the software calculations for the *proposed design*.

R405.5 Calculation procedure. Calculations of the performance design shall be in accordance with Sections R405.5.1 and R405.5.2.

R405.5.1 General. Except as specified by this section, the *standard reference design* and *proposed design* shall be configured and analyzed using identical methods and techniques.

R405.5.2 Residence specifications. The *standard reference design* and *proposed design* shall be configured and analyzed as specified by Table R405.5.2(1). Table R405.5.2(1) shall include, by reference, all notes contained in Table R402.1.2.

R405.6 Calculation software tools. Calculation software, where used, shall be in accordance with Sections R405.6.1 through R405.6.3.

R405.6.1 Minimum capabilities. Calculation procedures used to comply with this section shall be software tools capable of calculating the annual energy consumption of all building elements that differ between the *standard reference design* and the *proposed design* and shall include the following capabilities:

1. Computer generation of the *standard reference design* using only the input for the *proposed design*. The calculation procedure shall not allow the user to directly modify the building component characteristics of the *standard reference design*.
2. Calculation of whole-building (as a single *zone*) sizing for the heating and cooling equipment in the *standard reference design* residence in accordance with Section R403.6.
3. Calculations that account for the effects of indoor and outdoor temperatures and part-load ratios on the performance of heating, ventilating and air-conditioning equipment based on climate and equipment sizing.
4. Printed *code official* inspection checklist listing each of the *proposed design* component characteristics from Table R405.5.2(1) determined by the analysis to provide compliance, along with their respective performance ratings (*R*-value, *U*-factor, SHGC, HSPF, AFUE, SEER, EF are some examples).

R405.6.2 Specific approval. Performance analysis tools meeting the applicable provisions of Section R405 shall be permitted to be *approved*. Tools are permitted to be *approved* based on meeting a specified threshold for a jurisdiction. The *code official* shall be permitted to approve tools for a specified application or limited scope.

R405.6.3 Input values. When calculations require input values not specified by Sections R402, R403, R404 and R405, those input values shall be taken from an *approved* source.

TABLE R405.5.2(1)
SPECIFICATIONS FOR THE STANDARD REFERENCE AND PROPOSED DESIGNS

BUILDING COMPONENT	STANDARD REFERENCE DESIGN	PROPOSED DESIGN
Above-grade walls	Type: mass wall if proposed wall is mass; otherwise wood frame.	As proposed
	Gross area: same as proposed	As proposed
	U-factor: as specified in Table R402.1.4	As proposed
	Solar absorptance = 0.75	As proposed
	Emittance = 0.90	As proposed
Basement and crawl space walls	Type: same as proposed	As proposed
	Gross area: same as proposed	As proposed
	U-factor: from Table R402.1.4, with insulation layer on interior side of walls	As proposed
Above-grade floors	Type: wood frame	As proposed
	Gross area: same as proposed	As proposed
	U-factor: as specified in Table R402.1.4	As proposed
Ceilings	Type: wood frame	As proposed
	Gross area: same as proposed	As proposed
	U-factor: as specified in Table R402.1.4	As proposed
Roofs	Type: composition shingle on wood sheathing	As proposed
	Gross area: same as proposed	As proposed
	Solar absorptance = 0.75	As proposed
	Emittance = 0.90	As proposed
Attics	Type: vented with aperture = 1 ft^2 per 300 ft^2 ceiling area	As proposed
Foundations	Type: same as proposed	As proposed
	Foundation wall area above and below grade and soil characteristics: same as proposed	As proposed
Opaque doors	Area: 40 ft^2	As proposed
	Orientation: North	As proposed
	U-factor: same as fenestration from Table R402.1.4	As proposed
Vertical fenestration other than opaque doors	Total area[h] = (a) The proposed glazing area, where the proposed glazing area is less than 15 percent of the conditioned floor area (b) 15 percent of the conditioned floor area, where the proposed glazing area is 15 percent or more of the conditioned floor area	As proposed
	Orientation: equally distributed to four cardinal compass orientations (N, E, S & W).	As proposed
	U-factor: as specified in Table R402.1.4	As proposed
	SHGC: as specified in Table R402.1.2 except that for climates with no requirement (NR) SHGC = 0.40 shall be used.	As proposed
	Interior shade fraction: 0.92-(0.21 × SHGC for the standard reference design)	0.92-(0.21 × SHGC as proposed)
	External shading: none	As proposed
Skylights	None	As proposed
Thermally isolated sunrooms	None	As proposed

Air exchange rate	Air leakage rate of 5 air changes per hour in climate zones 1 and 2, and 3 air changes per hour in climate zones 3 through 8 at a pressure of 0.2 inches w.g (50 Pa). The mechanical ventilation rate shall be in addition to the air leakage rate and the same as in the proposed design, but no greater than $0.01 \times CFA + 7.5 \times (N_{br} + 1)$
	where:
	   CFA = conditioned floor area
	   N_{br} = number of bedrooms
	   Energy recovery shall not be assumed for mechanical ventilation.
		For residences that are not tested, the same air leakage rate as the standard reference design.
		
		For tested residences, the measured air exchange rate[a].
		
		The mechanical ventilation rate[b] shall be in addition to the air leakage rate and shall be as proposed.

(continued)

RESIDENTIAL ENERGY EFFICIENCY

TABLE R405.5.2(1)—continued
SPECIFICATIONS FOR THE STANDARD REFERENCE AND PROPOSED DESIGNS

BUILDING COMPONENT	STANDARD REFERENCE DESIGN	PROPOSED DESIGN
Mechanical ventilation	None, except where mechanical ventilation is specified by the proposed design, in which case: Annual vent fan energy use: $\quad$ kWh/yr = $0.03942 \times CFA + 29.565 \times (N_{br} + 1)$ where: $\quad CFA$ = conditioned floor area $\quad N_{br}$ = number of bedrooms	As proposed
Internal gains	IGain = $17{,}900 + 23.8 \times CFA + 4104 \times N_{br}$ (Btu/day per dwelling unit)	Same as standard reference design.
Internal mass	An internal mass for furniture and contents of 8 pounds per square foot of floor area.	Same as standard reference design, plus any additional mass specifically designed as a thermal storage element[c] but not integral to the building envelope or structure.
Structural mass	For masonry floor slabs, 80 percent of floor area covered by R-2 carpet and pad, and 20 percent of floor directly exposed to room air.	As proposed
	For masonry basement walls, as proposed, but with insulation required by Table R402.1.4 located on the interior side of the walls	As proposed
	For other walls, for ceilings, floors, and interior walls, wood frame construction	As proposed
Heating systems[d, e]	As proposed for other than electric heating without a heat pump, where the proposed design utilizes electric heating without a heat pump the standard reference design shall be an air source heat pump meeting the requirements of Section C403 of the IECC-Commercial Provisions. Capacity: sized in accordance with Section R403.7	As proposed
Cooling systems[d, f]	As proposed Capacity: sized in accordance with Section R403.7.	As proposed
Service water heating[d, e, f, g]	As proposed Use: same as proposed design	As proposed gal/day = $30 + (10 \times N_{br})$
Thermal distribution systems	Duct insulation: From Section R403.2.1 A thermal distribution system efficiency (DSE) of 0.88 shall be applied to both the heating and cooling system efficiencies for all systems other than tested duct systems. For tested duct systems, the leakage rate shall be 4 cfm (113.3 L/min) per 100 ft² (9.29 m²) of *conditioned floor* area at a pressure of differential of 0.1 inches w.g. (25 Pa).	As tested or as specified in Table R405.5.2(2) if not tested. Duct insulation shall be as proposed.
Thermostat	Type: Manual, cooling temperature setpoint = 75°F; Heating temperature setpoint = 72°F	Same as standard reference

For SI: 1 square foot = 0.93 m², 1 British thermal unit = 1055 J, 1 pound per square foot = 4.88 kg/m², 1 gallon (US) = 3.785 L,
$\quad$ °C = (°F-32)/1.8, 1 degree = 0.79 rad.

a. Where required by the *code official*, testing shall be conducted by an *approved* party. Hourly calculations as specified in the ASHRAE *Handbook of Fundamentals*, or the equivalent shall be used to determine the energy loads resulting from infiltration.
b. The combined air exchange rate for infiltration and mechanical ventilation shall be determined in accordance with Equation 43 of 2001 ASHRAE *Handbook of Fundamentals*, page 26.24 and the "Whole-house Ventilation" provisions of 2001 ASHRAE *Handbook of Fundamentals*, page 26.19 for intermittent mechanical ventilation.
c. Thermal storage element shall mean a component not part of the floors, walls or ceilings that is part of a passive solar system, and that provides thermal storage such as enclosed water columns, rock beds, or phase-change containers. A thermal storage element must be in the same room as fenestration that faces within 15 degrees (0.26 rad) of true south, or must be connected to such a room with pipes or ducts that allow the element to be actively charged.
d. For a proposed design with multiple heating, cooling or water heating systems using different fuel types, the applicable standard reference design system capacities and fuel types shall be weighted in accordance with their respective loads as calculated by accepted engineering practice for each equipment and fuel type present.
e. For a proposed design without a proposed heating system, a heating system with the prevailing federal minimum efficiency shall be assumed for both the standard reference design and proposed design.
f. For a proposed design home without a proposed cooling system, an electric air conditioner with the prevailing federal minimum efficiency shall be assumed for both the standard reference design and the proposed design.

(continued)

g. For a proposed design with a nonstorage-type water heater, a 40-gallon storage-type water heater with the prevailing federal minimum energy factor for the same fuel as the predominant heating fuel type shall be assumed. For the case of a proposed design without a proposed water heater, a 40-gallon storage-type water heater with the prevailing federal minimum efficiency for the same fuel as the predominant heating fuel type shall be assumed for both the proposed design and standard reference design.

h. For residences with conditioned basements, R-2 and R-4 residences and townhouses, the following formula shall be used to determine glazing area:
$AF = A_s \times FA \times F$
where:
AF = Total glazing area
A_s = Standard reference design total glazing area.
FA = (Above-grade thermal boundary gross wall area)/(above-grade boundary wall area + 0.5 x below-grade boundary wall area).
F = (Above-grade thermal boundary wall area)/(above-grade thermal boundary wall area + common wall area) or 0.56, whichever is greater.
and where:
Thermal boundary wall is any wall that separates conditioned space from unconditioned space or ambient conditions.
Above-grade thermal boundary wall is any thermal boundary wall component not in contact with soil.
Below-grade boundary wall is any thermal boundary wall in soil contact.
Common wall area is the area of walls shared with an adjoining dwelling unit.
L and CFA are in the same units.

TABLE R405.5.2(2)
DEFAULT DISTRIBUTION SYSTEM EFFICIENCIES FOR PROPOSED DESIGNS[a]

DISTRIBUTION SYSTEM CONFIGURATION AND CONDITION	FORCED AIR SYSTEMS	HYDRONIC SYSTEMS[b]
Distribution system components located in unconditioned space	—	0.95
Untested distribution systems entirely located in conditioned space[c]	0.88	1
"Ductless" systems[d]	1	—

For SI: 1 cubic foot per minute = 0.47 L/s, 1 square foot = 0.093 m², 1 pound per square inch = 6895 Pa, 1 inch water gauge = 1250 Pa.

a. Default values given by this table are for untested distribution systems, which must still meet minimum requirements for duct system insulation.
b. Hydronic systems shall mean those systems that distribute heating and cooling energy directly to individual spaces using liquids pumped through closed-loop piping and that do not depend on ducted, forced airflow to maintain space temperatures.
c. Entire system in conditioned space shall mean that no component of the distribution system, including the air-handler unit, is located outside of the conditioned space.
d. Ductless systems shall be allowed to have forced airflow across a coil but shall not have any ducted airflow external to the manufacturer's air-handler enclosure.

SECTION R406
ENERGY RATING INDEX
COMPLIANCE ALTERNATIVE

R406.1 Scope. This section establishes criteria for compliance using an Energy Rating Index (ERI) analysis.

R406.2 Mandatory requirements. Compliance with this section requires that the provisions identified in Sections R401 through R404 labeled as "mandatory" and Section R403.5.3 be met. The building thermal envelope shall be greater than or equal to levels of efficiency and Solar Heat Gain Coefficient in Table 402.1.1 or 402.1.3 of the 2009 *International Energy Conservation Code*.

> **Exception:** Supply and return ducts not completely inside the building thermal envelope shall be insulated to a minimum of R-6.

R406.3 Energy Rating Index. The Energy Rating Index (ERI) shall be a numerical integer value that is based on a linear scale constructed such that the *ERI reference design* has an Index value of 100 and a *residential building* that uses no net purchased energy has an Index value of 0. Each integer value on the scale shall represent a 1-percent change in the total energy use of the rated design relative to the total energy use of the *ERI reference design*. The ERI shall consider all energy used in the *residential building*.

R406.3.1 ERI reference design. The *ERI reference design* shall be configured such that it meets the minimum requirements of the 2006 *International Energy Conservation Code* prescriptive requirements.

The proposed *residential building* shall be shown to have an annual total normalized modified load less than or equal to the annual total loads of the *ERI reference design*.

R406.4 ERI-based compliance. Compliance based on an ERI analysis requires that the *rated design* be shown to have an ERI less than or equal to the appropriate value listed in Table R406.4 when compared to the *ERI reference design*.

R406.5 Verification by approved agency. Verification of compliance with Section R406 shall be completed by an *approved* third party.

TABLE R406.4
MAXIMUM ENERGY RATING INDEX

CLIMATE ZONE	ENERGY RATING INDEX
1	52
2	52
3	51
4	54
5	55
6	54
7	53
8	53

R406.6 Documentation. Documentation of the software used to determine the ERI and the parameters for the residential building shall be in accordance with Sections R406.6.1 through R406.6.3.

R406.6.1 Compliance software tools. Documentation verifying that the methods and accuracy of the compliance software tools conform to the provisions of this section shall be provided to the *code official*.

R406.6.2 Compliance report. Compliance software tools shall generate a report that documents that the ERI of the *rated design* complies with Sections R406.3 and R406.4. The compliance documentation shall include the following information:

1. Address or other identification of the residential building.
2. An inspection checklist documenting the building component characteristics of the *rated design*. The inspection checklist shall show results for both the *ERI reference design* and the *rated design*, and shall document all inputs entered by the user necessary to reproduce the results.
3. Name of individual completing the compliance report.
4. Name and version of the compliance software tool.

Exception: Multiple orientations. Where an otherwise identical building model is offered in multiple orientations, compliance for any orientation shall be permitted by documenting that the building meets the performance requirements in each of the four (north, east, south and west) cardinal orientations.

R406.6.3 Additional documentation. The *code official* shall be permitted to require the following documents:

1. Documentation of the building component characteristics of the *ERI reference design*.
2. A certification signed by the builder providing the building component characteristics of the *rated design*.
3. Documentation of the actual values used in the software calculations for the *rated design*.

R406.7 Calculation software tools. Calculation software, where used, shall be in accordance with Sections R406.7.1 through R406.7.3.

R406.7.1 Minimum capabilities. Calculation procedures used to comply with this section shall be software tools capable of calculating the ERI as described in Section R406.3, and shall include the following capabilities:

1. Computer generation of the *ERI reference design* using only the input for the *rated design*.

    The calculation procedure shall not allow the user to directly modify the building component characteristics of the *ERI reference design*.

2. Calculation of whole building, as a single *zone*, sizing for the heating and cooling equipment in the *ERI reference design* residence in accordance with Section R403.7.

3. Calculations that account for the effects of indoor and outdoor temperatures and part-load ratios on the performance of heating, ventilating and air-conditioning equipment based on climate and equipment sizing.

4. Printed *code official* inspection checklist listing each of the *rated design* component characteristics determined by the analysis to provide compliance, along with their respective performance ratings.

R406.7.2 Specific approval. Performance analysis tools meeting the applicable sections of Section R406 shall be *approved*. Tools are permitted to be *approved* based on meeting a specified threshold for a jurisdiction. The *code official* shall approve tools for a specified application or limited scope.

R406.7.3 Input values. When calculations require input values not specified by Sections R402, R403, R404 and R405, those input values shall be taken from an approved source.

CHAPTER 5 [RE]
EXISTING BUILDINGS

SECTION R501
GENERAL

R501.1 Scope. The provisions of this chapter shall control the *alteration*, repair, addition and change of occupancy of existing buildings and structures.

R501.1.1 Additions, alterations, or repairs: General. Additions, alterations, or repairs to an existing building, building system or portion thereof shall comply with Section R502, R503 or R504. Unaltered portions of the existing building or building supply system shall not be required to comply with this code.

R501.2 Existing buildings. Except as specified in this chapter, this code shall not be used to require the removal, *alteration* or abandonment of, nor prevent the continued use and maintenance of, an existing building or building system lawfully in existence at the time of adoption of this code.

R501.3 Maintenance. Buildings and structures, and parts thereof, shall be maintained in a safe and sanitary condition. Devices and systems that are required by this code shall be maintained in conformance to the code edition under which installed. The owner or the owner's authorized agent shall be responsible for the maintenance of buildings and structures. The requirements of this chapter shall not provide the basis for removal or abrogation of energy conservation, fire protection and safety systems and devices in existing structures.

R501.4 Compliance. *Alterations*, *repairs*, *additions* and changes of occupancy to, or relocation of, existing buildings and structures shall comply with the provisions for *alterations*, *repairs*, *additions* and changes of occupancy or relocation, respectively, in the *International Residential Code*, *International Building Code*, *International Fire Code*, *International Fuel Gas Code*, *International Mechanical Code*, *International Plumbing Code*, *International Property Maintenance Code*, *International Private Sewage Disposal Code* and NFPA 70.

R501.5 New and replacement materials. Except as otherwise required or permitted by this code, materials permitted by the applicable code for new construction shall be used. Like materials shall be permitted for repairs, provided hazards to life, health or property are not created. Hazardous materials shall not be used where the code for new construction would not permit their use in buildings of similar occupancy, purpose and location.

R501.6 Historic buildings. No provision of this code relating to the construction, *repair*, *alteration*, restoration and movement of structures, and *change of occupancy* shall be mandatory for *historic buildings* provided a report has been submitted to the code official and signed by the owner, a registered *design professional*, or a representative of the State Historic Preservation Office or the historic preservation authority having jurisdiction, demonstrating that compliance with that provision would threaten, degrade or destroy the historic form, fabric or function of the *building*.

SECTION R502
ADDITIONS

R502.1 General. Additions to an existing building, building system or portion thereof shall conform to the provisions of this code as those provisions relate to new construction without requiring the unaltered portion of the existing building or building system to comply with this code. Additions shall not create an unsafe or hazardous condition or overload existing building systems. An addition shall be deemed to comply with this code where the addition alone complies, where the existing building and addition comply with this code as a single building, or where the building with the addition uses no more energy than the existing building. Additions shall be in accordance with Section R502.1.1 or R502.1.2.

R502.1.1 Prescriptive compliance. Additions shall comply with Sections R502.1.1.1 through R502.1.1.4.

R502.1.1.1 Building envelope. New building envelope assemblies that are part of the addition shall comply with Sections R402.1, R402.2, R402.3.1 through R402.3.5, and R402.4.

Exception: Where nonconditioned space is changed to conditioned space, the building envelope of the addition shall comply where the UA, as determined in Section 402.1.5, of the existing building and the addition, and any alterations that are part of the project, is less than or equal to UA generated for the existing building.

R502.1.1.2 Heating and cooling systems. New heating, cooling and duct systems that are part of the addition shall comply with Sections R403.1, R403.2, R403.3, R403.5 and R403.6.

Exception: Where ducts from an existing heating and cooling system are extended to an addition, duct systems with less than 40 linear feet (12.19 m) in unconditioned spaces shall not be required to be tested in accordance with Section R403.3.3.

R502.1.1.3 Service hot water systems. New service hot water systems that are part of the addition shall comply with Section R403.4.

R502.1.1.4 Lighting. New lighting systems that are part of the addition shall comply with Section R404.1.

R502.1.2 Existing plus addition compliance (Simulated Performance Alternative). Where nonconditioned space is changed to conditioned space, the addition shall comply where the annual energy cost or energy use of the addition and the existing building, and any alterations that are part of the project, is less than or equal to the annual energy cost of the existing building when modeled in accordance with Section R405. The addition and any alterations that are part of the project shall comply with Section R405 in its entirety.

SECTION R503
ALTERATIONS

R503.1 General. *Alterations* to any building or structure shall comply with the requirements of the code for new construction. *Alterations* shall be such that the existing building or structure is no less conforming to the provisions of this code than the existing building or structure was prior to the *alteration*.

Alterations to an existing building, building system or portion thereof shall conform to the provisions of this code as they relate to new construction without requiring the unaltered portions of the existing building or building system to comply with this code. Alterations shall not create an unsafe or hazardous condition or overload existing building systems. *Alterations* shall be such that the existing building or structure uses no more energy than the existing building or structure prior to the *alteration*. Alterations to existing buildings shall comply with Sections R503.1.1 through R503.2.

R503.1.1 Building envelope. Building envelope assemblies that are part of the alteration shall comply with Section R402.1.2 or R402.1.4, Sections R402.2.1 through R402.2.13, R402.3.1, R402.3.2, R402.4.3 and R402.4.4.

> **Exception:** The following alterations need not comply with the requirements for new construction provided the energy use of the building is not increased:
> 1. Storm windows installed over existing fenestration.
> 2. Existing ceiling, wall or floor cavities exposed during construction provided that these cavities are filled with insulation.
> 3. Construction where the existing roof, wall or floor cavity is not exposed.
> 4. Roof recover.
> 5. Roofs without insulation in the cavity and where the sheathing or insulation is exposed during reroofing shall be insulated either above or below the sheathing.
> 6. Surface-applied window film installed on existing single pane fenestration assemblies to reduce solar heat gain provided the code does not require the glazing or fenestration assembly to be replaced.

R503.1.1.1 Replacement fenestration. Where some or all of an existing fenestration unit is replaced with a new fenestration product, including sash and glazing, the replacement fenestration unit shall meet the applicable requirements for *U*-factor and SHGC as provided in Table R402.1.2.

R503.1.2 Heating and cooling systems. New heating, cooling and duct systems that are part of the alteration shall comply with Sections R403.1, R403.2, R403.3 and R403.6.

> **Exception:** Where ducts from an existing heating and cooling system are extended, duct systems with less than 40 linear feet (12.19 m) in unconditioned spaces shall not be required to be tested in accordance with Section R403.3.3.

R503.1.3 Service hot water systems. New service hot water systems that are part of the alteration shall comply with Section R403.4.

R503.1.4 Lighting. New lighting systems that are part of the alteration shall comply with Section 404.1.

> **Exception:** Alterations that replace less than 50 percent of the luminaires in a space, provided that such alterations do not increase the installed interior lighting power.

R503.2 Change in space conditioning. Any nonconditioned or low-energy space that is altered to become *conditioned space* shall be required to be brought into full compliance with this code.

> **Exception:** Where the simulated performance option in Section R405 is used to comply with this section, the annual energy cost of the proposed design is permitted to be 110 percent of the annual energy cost otherwise allowed by Section R405.3.

SECTION R504
REPAIRS

R504.1 General. Buildings, structures and parts thereof shall be repaired in compliance with Section R501.3 and this section. Work on nondamaged components necessary for the required *repair* of damaged components shall be considered part of the *repair* and shall not be subject to the requirements for *alterations* in this chapter. Routine maintenance required by Section R501.3, ordinary repairs exempt from *permit*, and abatement of wear due to normal service conditions shall not be subject to the requirements for *repairs* in this section.

R504.2 Application. For the purposes of this code, the following shall be considered repairs:

1. Glass-only replacements in an existing sash and frame.
2. Roof repairs.
3. Repairs where only the bulb and/or ballast within the existing luminaires in a space are replaced provided that the replacement does not increase the installed interior lighting power.

SECTION R505
CHANGE OF OCCUPANCY OR USE

R505.1 General. Spaces undergoing a change in occupancy that would result in an increase in demand for either fossil fuel or electrical energy shall comply with this code.

R505.2 General. Any space that is converted to a dwelling unit or portion thereof from another use or occupancy shall comply with this code.

> **Exception:** Where the simulated performance option in Section R405 is used to comply with this section, the annual energy cost of the proposed design is permitted to be 110 percent of the annual energy cost otherwise allowed by Section R405.3.

CHAPTER 6 [RE]
REFERENCED STANDARDS

This chapter lists the standards that are referenced in various sections of this document. The standards are listed herein by the promulgating agency of the standard, the standard identification, the effective date and title, and the section or sections of this document that reference the standard. The application of the referenced standards shall be as specified in Section 106.

AAMA

American Architectural Manufacturers Association
1827 Walden Office Square
Suite 550
Schaumburg, IL 60173-4268

Standard reference number	Title	Referenced in code section number
AAMA/WDMA/CSA 101/I.S.2/A C440—11	North American Fenestration Standard/ Specifications for Windows, Doors and Unit Skylights	R402.4.3

ACCA

Air Conditioning Contractors of America
2800 Shirlington Road, Suite 300
Arlington, VA 22206

Standard reference number	Title	Referenced in code section number
Manual J—2011	Residential Load Calculation Eighth Edition	R403.7
Manual S—13	Residential Equipment Selection	R403.7

APSP

The Association of Pool and Spa Professionals
2111 Eisenhower Avenue
Alexandria, VA 22314

Standard reference number	Title	Referenced in code section number
APSP 14—11	American National Standard for Portable Electric Spa Energy Efficiency	R403.10.1, 403.11
APSP 15a—2013	American National Standard for Residential Swimming Pool and Spa Energy Efficiency	R403.12

ASHRAE

American Society of Heating, Refrigerating and Air-Conditioning Engineers, Inc.
1791 Tullie Circle, NE
Atlanta, GA 30329-2305

Standard reference number	Title	Referenced in code section number
ASHRAE—2013	ASHRAE Handbook of Fundamentals	R402.1.5, Table R405.5.2(1)
ASHRAE 193—2010	Method of Test for Determining the Airtightness of HVAC Equipment	R403.3.2.1

REFERENCED STANDARDS

ASTM

ASTM International
100 Barr Harbor Drive
West Conshohocken, PA 19428-2859

Standard reference number	Title	Referenced in code section number
C 1363—11	Standard Test Method for Thermal Performance of Building Materials and Envelope Assemblies by Means of a Hot Box Apparatus	R303.1.4.1
E 283—04	Test Method for Determining the Rate of Air Leakage Through Exterior Windows, Curtain Walls and Doors Under Specified Pressure Differences Across the Specimen	R402.4.4
E 779—10	Standard Test Method for Determining Air Leakage Rate by Fan Pressurization	R402.4.1.2
E 1827—11	Standard Test Methods for Determining Airtightness of Building Using an Orifice Blower Door	R402.4.1.2

CSA

CSA Group
8501 East Pleasant Valley
Cleveland, OH 44131-5575

Standard reference number	Title	Referenced in code section number
AAMA/WDMA/CSA 101/I.S.2/A440—11	North American Fenestration Standard/Specification for Windows, Doors and Unit Skylights	R402.4.3
CSA 55.1—2012	Test Method for measuring efficiency and pressure loss of drain water heat recovery units	R403.5.4
CSA 55.2—2012	Drain water heat recover units	R403.5.4

DASMA

Door and Access Systems Manufacturers Association
1300 Sumner Avenue
Cleveland, OH 44115-2851

Standard reference number	Title	Referenced in code section number
105—92(R2004)—13	Test Method for Thermal Transmittance and Air Infiltration of Garage Doors	R303.1.3

ICC

International Code Council, Inc.
500 New Jersey Avenue, NW
6th Floor
Washington, DC 20001

Standard reference number	Title	Referenced in code section number
IBC—15	International Building Code®	R201.3, R303.2, R402.1.1, R501.4
ICC 400—12	Standard on the Design and Construction of Log Structures	Table R402.5.1.1
IECC—15	International Energy Conservation Code®	R101.4.1, 403.8
IECC—09	2009 International Energy Conservation Code®	R406.2
ECC—06	2006 International Energy Conservation Code®	R202, R406.3.1
IFC—15	International Fire Code®	R201.3, R501.4
IFGC—15	International Fuel Gas Code®	R201.3, R501.4
IMC—15	International Mechanical Code®	R201.3, R403.3.2, R403.6, R501.4
IPC—15	International Plumbing Code®	R201.3, R501.4
IPSDC—15	International Private Sewage Disposal Code®	501.4
IPMC—15	International Property Maintenance Code®	501.4
IRC—15	International Residential Code®	R201.3, R303.2, R402.1.1, R402.2.11, R403.3.2, R403.6, R501.4

REFERENCED STANDARDS

IEEE

The Institute of Electrical and Electronic Engineers, Inc.
3 Park Avenue
New York, NY 1016-5997

Standard reference number	Title	Referenced in code section number
515.1—2012	IEEE Standard for the Testing, Design, Installation, and Maintenance of Electrical Resistance Trace Heating for Commercial Applications	R403.5.1.2

NFPA

National Fire Protection Association.
1 Batterymarch Park
Quincy, MA 02169-7471

Standard reference number	Title	Referenced in code section number
70—14	National Electrical Code	R501.4

NFRC

National Fenestration Rating Council, Inc.
6305 Ivy Lane, Suite 140
Greenbelt, MD 20770

Standard reference number	Title	Referenced in code section number
100—2009	Procedure for Determining Fenestration Products U-factors—Second Edition	R303.1.3
200—2009	Procedure for Determining Fenestration Product Solar Heat Gain Coefficients and Visible Transmittance at Normal Incidence—Second Edition	R303.1.3
400—2009	Procedure for Determining Fenestration Product Air Leakage—Second Edition	R402.4.3

UL

UL LLC
333 Pfingsten Road
Northbrook, IL 60062

Standard reference number	Title	Referenced in code section number
127—11	Standard for Factory Built Fireplaces	R402.4.2
515—11	Electrical Resistance Heat Tracing for Commercial and Industrial Applications including revisions through November 30, 2011	R403.5.1.2

US-FTC

United States-Federal Trade Commission
600 Pennsylvania Avenue NW
Washington, DC 20580

Standard reference number	Title	Referenced in code section number
CFR Title 16 (May 31, 2005)	R-value Rule	R303.1.4

REFERENCED STANDARDS

Window and Door Manufacturers Association
2025 M Street, NW Suite 800
Washington, DC 20036-3309

Standard reference number	Title	Referenced in code section number
AAMA/WDMA/CSA 101/I.S.2/A440—11	North American Fenestration Standard/Specification for Windows, Doors and Unit Skylights.	R402.4.3

APPENDIX RA

RECOMMENDED PROCEDURE FOR WORST-CASE TESTING OF ATMOSPHERIC VENTING SYSTEMS UNDER R402.4 OR R405 CONDITIONS ≤ 5ACH$_{50}$

(This appendix is informative and is not part of the code.)

SECTION RA101
SCOPE

RA101.1 General. This appendix is intended to provide guidelines for worst-case testing of atmospheric venting systems. Worst-case testing is recommended to identify problems that weaken draft and restrict combustion air.

SECTION RA201
GENERAL DEFINITIONS

COMBUSTION APPLIANCE ZONE (CAZ). A contiguous air volume within a building that contains a Category I or II atmospherically vented appliance or a Category III or IV direct-vent or integral vent appliance drawing combustion air from inside the building or dwelling unit. The CAZ includes, but is not limited to, a mechanical closet, a mechanical room, or the main body of a house or dwelling unit.

DRAFT. The pressure difference existing between the *appliance* or any component part and the atmosphere that causes a continuous flow of air and products of *combustion* through the gas passages of the *appliance* to the atmosphere.

    Mechanical or induced draft. The pressure difference created by the action of a fan, blower or ejector that is located between the *appliance* and the *chimney* or vent termination.

    Natural draft. The pressure difference created by a vent or *chimney* because of its height and the temperature difference between the *flue* gases and the atmosphere.

SPILLAGE. Combustion gases emerging from an appliance or venting system into the combustion appliance zone during burner operation.

SECTION RA301
TESTING PROCEDURE

RA301.1 Worst-case testing of atmospheric venting systems. Buildings or dwelling units containing a Category I or II atmospherically vented appliance; or a Category III or IV direct-vent or integral vent appliance drawing combustion air from inside of the building or dwelling unit, shall have the Combustion Appliance Zone (CAZ) tested for spillage, acceptable draft and carbon monoxide (CO) in accordance with this section. Where required by the *code official*, testing shall be conducted by an *approved* third party. A written report of the results of the test shall be signed by the party conducting the test and provided to the *code official*. Testing shall be performed at any time after creation of all penetrations of the *building thermal* envelope and prior to final inspection.

    Exception: Buildings or dwelling units containing only Category III or IV direct-vent or integral vent appliances that do not draw combustion air from inside of the building or dwelling unit.

The enumerated test procedure as follows shall be complied with during testing:

1. Set combustion appliances to the pilot setting or turn off the service disconnects for combustion appliances. Close exterior doors and windows and the fireplace damper. With the building or dwelling unit in this configuration, measure and record the baseline ambient pressure inside the building or dwelling unit CAZ. Compare the baseline ambient pressure of the CAZ to that of the outside ambient pressure and record the difference (Pa).

2. Establish worst case by turning on the *clothes dryer* and all exhaust fans. Close all interior doors that make the CAZ pressure more negative. Turn on the air handler, where present, and leave on if, as a result, the pressure in the CAZ becomes more negative. Check interior door positions again, closing only the interior doors that make the CAZ pressure more negative. Measure net change in pressure from the CAZ to outdoor ambient pressure, correcting for the base ambient pressure inside the home. Record "worst case depressurization" pressure and compare to Table RA301.1(1).

   Where CAZ depressurization limits are exceeded under worst-case conditions in accordance with Table A301.1(1), additional combustion air shall be provided or other modifications to building air-leakage performance or exhaust appliances such that depressurization is brought within the limits prescribed in Table RA301.1(1).

3. Measure worst-case spillage, acceptable draft and carbon monoxide (CO) by firing the fuel-fired appliance with the smallest Btu capacity first.

   a. Test for spillage at the draft diverter with a mirror or smoke puffer. An appliance that continues to spill flue gases for more than 60 seconds fails the spillage test.

   b. Test for CO measuring undiluted flue gases in the throat or flue of the appliance using a digital gauge in parts per million (ppm) at the 10-minute mark. Record CO ppm readings to be compared with Table

RA301.1(3) upon completion of Step 4. Where the spillage test fails under worst case, go to Step 4.

c. Where spillage ends within 60 seconds, test for acceptable draft in the connector not less than 1 foot (305 mm), but not more than 2 feet (610 mm) downstream of the draft diverter. Record draft pressure and compare to Table RA301.1(2).

d. Fire all other connected appliances simultaneously and test again at the draft diverter of each appliance for spillage, CO and acceptable draft using procedures 3a through 3c.

4. Measure spillage, acceptable draft, and carbon monoxide (CO) under natural conditions—without *clothes dryer* and exhaust fans on—in accordance with the procedure outlined in Step 3, measuring the net change in pressure from worst case condition in Step 3 to natural in the CAZ to confirm the worst case depressurization taken in Step 2. Repeat the process for each appliance, allowing each vent system to cool between tests.

5. Monitor indoor ambient CO in the breathing zone continuously during testing, and abort the test where indoor ambient CO exceeds 35 ppm by turning off the appliance, ventilating the space, and evacuating the building. The CO problem shall be corrected prior to completing combustion safety diagnostics.

6. Make recommendations based on test results and the retrofit action prescribed in Table RA301.1(3).

TABLE RA301.1(1)
CAZ DEPRESSURIZATION LIMITS

VENTING CONDITION	LIMIT (Pa)
Category I, atmospherically vented water heater	−2.0
Category I or II atmospherically vented boiler or furnace common-vented with a Category I atmospherically vented water heater	−3.0
Category I or II atmospherically vented boiler or furnace, equipped with a flue damper, and common vented with a Category I atmospherically vented water heater	−5.0
Category I or II atmospherically vented boiler or furnace alone	
Category I or II atmospherically vented, fan-assisted boiler or furnace common vented with a Category I atmospherically vented water heater	
Decorative vented, gas appliance	
Power-vented or induced-draft boiler or furnace alone, or fan-assisted water heater alone	−15.0
Category IV direct-vented appliances and sealed combustion appliances	−50.0

For SI: 6894.76 Pa = 1.0 psi.

TABLE RA301.1(2)
ACCEPTABLE DRAFT TEST CORRECTION

OUTSIDE TEMPERATURE (°F)	MINIMUM DRAFT PRESSURE REQUIRED (Pa)
< 10	−2.5
10 – 90	(Outside Temperature ÷ 40) − 2.75
> 90	−0.5

For SI: 6894.76 Pa = 1.0 psi.

TABLE RA301.1(3)
ACCEPTABLE DRAFT TEST CORRECTION

CARBON MONOXIDE LEVEL (ppm)	AND OR	SPILLAGE AND ACCEPTABLE DRAFT TEST RESULTS	RETROFIT ACTION
0 – 25	and	Passes	Proceed with work
25 < x ≤ 100	and	Passes	Recommend that CO problem be resolved
25 < x ≤ 100	and	Fails in worst case only	Recommend an appliance service call and repairs to resolve the problem
100 < x ≤ 400	or	Fails under natural conditions	**Stop!** Work shall not proceed until appliance is serviced and problem resolved
> 400	and	Passes	**Stop!** Work shall not proceed until appliance is serviced and problem resolved
> 400	and	Fails under any condition	**Emergency!** Shut off fuel to appliance and call for service immediately

APPENDIX RB

SOLAR-READY PROVISIONS—DETACHED ONE- AND TWO-FAMILY DWELLINGS, MULTIPLE SINGLE-FAMILY DWELLINGS (TOWNHOUSES)

(The provisions contained in this appendix are not mandatory unless specifically referenced in the adopting ordinance.)

SECTION RB101
SCOPE

RB101.1 General. These provisions shall be applicable for new construction where solar-ready provisions are required.

SECTION RB102
GENERAL DEFINITION

SOLAR-READY ZONE. A section or sections of the roof or building overhang designated and reserved for the future installation of a solar photovoltaic or solar thermal system.

SECTION RB103
SOLAR-READY ZONE

RB103.1 General. New detached one- and two-family dwellings, and multiple single-family dwellings (townhouses) with not less than 600 square feet (55.74 m^2) of roof area oriented between 110 degrees and 270 degrees of true north shall comply with Sections RB103.2 through RB103.8.

Exceptions:

1. New residential buildings with a permanently installed on-site renewable energy system.

2. A building with a solar-ready zone that is shaded for more than 70 percent of daylight hours annually.

RB103.2 Construction document requirements for solar-ready zone. Construction documents shall indicate the solar-ready zone.

RB103.3 Solar-ready zone area. The total solar-ready zone area shall be not less than 300 square feet (27.87 m^2) exclusive of mandatory access or set back areas as required by the *International Fire Code*. New multiple single-family dwellings (townhouses) three stories or less in height above grade plane and with a total floor area less than or equal to 2,000 square feet (185.8 m^2) per dwelling shall have a solar-ready zone area of not less than 150 square feet (13.94 m^2). The solar-ready zone shall be composed of areas not less than 5 feet (1524 mm) in width and not less than 80 square feet (7.44 m^2) exclusive of access or set back areas as required by the *International Fire Code*.

RB103.4 Obstructions. Solar-ready zones shall be free from obstructions, including but not limited to vents, chimneys, and roof-mounted equipment.

RB103.5 Roof load documentation. The structural design loads for roof dead load and roof live load shall be clearly indicated on the construction documents.

RB103.6 Interconnection pathway. Construction documents shall indicate pathways for routing of conduit or plumbing from the solar-ready zone to the electrical service panel or service hot water system.

RB103.7 Electrical service reserved space. The main electrical service panel shall have a reserved space to allow installation of a dual pole circuit breaker for future solar electric installation and shall be labeled "For Future Solar Electric." The reserved space shall be positioned at the opposite (load) end from the input feeder location or main circuit location.

RB103.8 Construction documentation certificate. A permanent certificate, indicating the solar-ready zone and other requirements of this section, shall be posted near the electrical distribution panel, water heater or other conspicuous location by the builder or registered design professional.

INDEX

A

ACCESS HATCHES R402.2.4
ADDITIONS
  Defined R202
  Requirements R501.1.1, R502
ALTERATIONS
  Defined R202
  Requirements 503
ADMINISTRATION . . . . Chapter 1
AIR BARRIER
  Installation . . . . R402.4.1.1, Table R402.4.1.1
  Testing . . . . ???
AIR ECONOMIZERS
  Defined R202
AIR INFILTRATION
  Defined R202
  Requirements R402.8
AIR LEAKAGE R402.4, R403.3.3, R403.3.4
ALTERNATE MATERIALS R102
APPROVED
  Defined R202
AUTOMATIC
  Defined R202
ATMOSPHERIC VENTING . . . . Appendix RA
  Worst case testing RA103

B

BASEMENT WALLS
  Defined R202
  Requirements . . . . R303.2.1, Table R402.1.2,
                            Table R402.1.3, R402.2.8,
                                  Table R405.5.2(1)
BELOW-GRADE WALLS (see BASEMENT WALLS)
BOARD OF APPEALS R109
  Qualifications of members R109.3
BUILDING
  Defined R202
BUILDING THERMAL ENVELOPE
  Air tightness R402.4.1
  Compliance documentation R103.2, R401.2
  Defined R202
  Insulation R303.1.1
  Insulation and fenestration criteria R402.1.2
  Performance method . . . . Table R405.5.2(1)
  Requirements R102.1.1, R402

C

CEILINGS R402.2.1, R402.2.2
  Specification for standard
    reference design . . . . Table R405.5.2(1)
CERTIFICATE R401.3, B103.8
CHANGE OF OCCUPANCY R505
CIRCULATON SYSTEMS R403.5.1
CLIMATE TYPES
  Defined . . . . Table R301.3(1)
CLIMATE ZONES . . . . R301, Figure R301.1,
                                        Table R301.1
  By state or territory . . . . Figure R301.1, Table R301.1
  International climate zones R301.3,
                   Table R301.3(1), Table R301.3(2)
  Tropical R301.4
  Warm humid . . . . . . R301.2, Table R301.2, R301.3.1
CODE OFFICIAL
  Defined R202
COMMERCIAL BUILDINGS
  Compliance R101.5
  Defined R202
COMPLIANCE AND ENFORCEMENT R101.5
  Compliance Report R405.4.2.2
CONDITIONED FLOOR AREA
  Defined R202
CONDITIONED SPACE
  Defined R202
CONSTRUCTION DOCUMENTS R103
  Amended R103.4
  Approval R103.3.1
  Examination R103.3
  Information required R103.2
  Phased R103.3.3
  Previous R103.3.2
  Retention R103.5
  Thermal envelope depiction R405.4.2.2
CONTINUOUS AIR BARRIER
  Defined R202
CONTROLS
  Heat pump R403.1.1
  Heating and cooling R403.1
  Service water heating R403.5
CRAWL SPACE WALLS
  Defined R202
  Requirements . . . . R303.2.1, Table R402.1.2,
                          Table R402.1.3, R402.1.4,
                          R402.2.10, Table R405.5.2(1)

INDEX

D

DEFAULT DISTRIBUTION SYSTEM EFFICIENCIES .... Table R405.5.2(2)
DEFINITIONS .... Chapter 2
DEGREE DAY COOLING .... Table R301.3(2)
DEGREE DAY HEATING .... Table R301.3(2)
DEMAND RECIRCULATION WATER SYSTEM
   Defined R202
   Requirements R403.5.2
DESIGN CONDITIONS .... Chapter 3, R302
DOORS
   Attics and crawl spaces R402.2.4
   Default U-factors .... Table R303.1.3(2)
   Opaque R402.3.4
   Performance requirements ...... Table R405.5.2(1)
   SHGC values .... Table R402.1.2
   U-values .... Table R402.1.4
DUCTS
   Defined R202
   Insulation R103.2, R401.3, R403.3, R403.3.1
   Sealing R103.2, R403.3.2
   Tightness verification
      Postconstruction test R403.3.2
      Rough-in test R403.3.2
DUCT SYSTEM
   Defined R202
DWELLING UNIT
   Defined R202
   Multiple units R403.8
DYNAMIC GLAZING R402.3.2
   Defined R202

E

EAVE BAFFLE
   Installation R402.2.3
ELECTRICAL POWER AND LIGHTING R404
ENERGY ANALYSIS, ANNUAL
   Defined R202
   Documentation R405.4
   Requirements R405.3
ENERGY COST
   Defined R202
   Energy rating index compliance alternative R406
   Energy rating index R406.3
   ERI-based compliance R406.4
ENERGY RECOVERY VENTILATION SYSTEMS
   Defined R202
   Requirements .... Table R405.5.2(1)

ENERGY SIMULATION TOOL
   Defined R202
ENTRANCE DOOR
   Defined R202
ENVELOPE, BUILDING THERMAL
   Defined R202
ENVELOPE DESIGN PROCEDURES R402
EQUIPMENT EFFICIENCIES R103.2, R401.3
EQUIPMENT ROOM
   Defined R202
EXISTING BUILDINGS .... Chapter 5
EXTERIOR WALLS
   Defined R202
   Thermal performance .... R402, R402.1.2, Table R405.5.2(1)

F

FEES R107
   Refunds R107.5
   Related fees R107.4
   Schedule of permit fees R107.2
FENESTRATION R303.1.3, R402.3, R402.3.2, R402.4.3
   Default U-factors .... Table R303.1.3(1)
   Defined R202
   Rating and labeling R303.1.3, R402.1.1
   Replacement R402.3.6
   Requirements .... Table R402.1.2
FENESTRATION PRODUCT, FIELD-FABRICATED
   Defined R202
FENESTRATION PRODUCT, SITE-BUILT
   Defined R202
FIREPLACES R402.4.2
FLOORS
   Above-grade .... Table R405.5.2(1)
   Insulation R402.2.7
   Slab-on-grade insulation requirements R402.2.9
FOUNDATIONS
   Requirements .... Table R402.5.1.1, Table R405.5.2(1)
FURNACE EFFICIENCY .... Table R405.5.2(1)

G

GLAZED FENESTRATION R402.3.2, R402.3.3

INDEX

H

HEAT PUMP R403.1.2
HEAT TRAPS
  Defined R202
HEATED SLAB
  Defined R202
HEATING AND COOLING LOADS .. R302.1, R403.1.2
HIGH-EFFICACY LAMPS
  Defined R202
HISTORIC BUILDINGS R101.4.2
HOT WATER
  Piping insulation .... R403.4.2, Table R403.4.2
HOT WATER BOILER
  Outdoor temperature setback R403.2
HVAC SYSTEMS
  Tests
    Postconstruction R403.3.4
    Rough-in-test R403.3.4

I

IDENTIFICATION (MATERIALS, EQUIPMENT AND SYSTEM) R303.1
INDIRECTLY CONDITIONED SPACE
  (see CONDITIONED SPACE)
INFILTRATION, AIR LEAKAGE R402.4,
  Table R405.5.2(1)
  Defined R202
INSPECTIONS R104
INSULATION
  Basement walls R402.2.9
  Ceilings with attic spaces R402.2.1
  Ceilings without attic spaces R402.2.2
  Crawl space walls.................... R402.2.10
  Duct R403.2.1
  Eave baffle R402.2.3
  Floors R402.2.8
  Hot water piping R403.4.2
  Identification R303.1, R303.1.2
  Installation R303.1.1, R303.1.1.1,
      R303.1.2, R303.2, Table R402.4.1.2
  Masonry veneer R402.2.11
  Mass walls R402.2.5
  Mechanical system piping R403.3
  Product rating R303.1.4
  Protection of exposed foundation R303.2.1
  Protection of piping insulation........... R403.3.1
  Requirements .... Table R402.1.1, R402.2
  Slab-on-grade floors R402.2.9
  Steel-frame ceilings, walls and floors R402.2.6,
      Table R402.2.6
  Sunroom R402.2.12
INSULATING SHEATHING
  Defined R202
  Requirements .... Table R402.1.2, R402.1.2

L

LABELED
  Defined R202
  Requirements R303.1.3, R303.3, R402.4.3
LIGHTING POWER
  Manufacturer's information R102.2, R102.3
LIGHTING SYSTEMS R404
  Recessed R402.4.4, R404
LISTED
  Defined R202
LOW ENERGY BUILDINGS R101.5.2
LOW-VOLTAGE LIGHTING
  Defined R202
LUMINAIRE
  Sealed.............................. R402.4.3

M

MAINTENANCE INFORMATION R102.3
MANUALS R101.5.1, R102.3
  Defined R202
MASONRY VENEER
  Insulation........................... R402.2.12
MASS
  Wall...................... 402.1.2, R402.1.3,
      R402.2.5, Table R402.2.6
MATERIALS AND EQUIPMENT R303
MECHANICAL SYSTEMS AND EQUIPMENT..................... R403, R405.1
MECHANICAL VENTILATION R403.6,
  Table R403.5.1, Table R405.5.2(1)
MULTIPLE DWELLING UNITS R403.8

O

OCCUPANCY
  Requirements R101.4.4, R101.4.6, R101.5
OPAQUE DOORS R402.3.4

P

PERFORMANCE ANALYSIS R405
PERMIT
  (see FEES)
  Work commencing before permit R107.3
PIPE INSULATION R403.4, R403.5

INDEX

PLANS AND SPECIFICATIONS R103
POOLS R403.10
   Covers R403.10.4
   Heaters R403.10.2
   Time switches R403.10.3
PROPOSED DESIGN
   Defined R202
   Requirements . . . . R405, Table R405.5.2(1)
PUMPS
   Time switches R403.10.3

R

R-VALUE
   Defined R202
   Computation R402.1.3
   Wood frame to cold formed
      steel frame ceiling, wall and
      floor insulation R-values . . . . Table R402.2.6
READILY ACCESSIBLE
   Defined R202
REFERENCED STANDARDS . . . . . . R106, Chapter 6
REPAIR
   Defined R202
   Requirements R504
RESIDENTIAL BUILDINGS
   Compliance R101.2, R101.5
   Defined R202
   Simulated performance alternative R405
ROOF ASSEMBLY
   Defined R202
   Requirements R101.4.3, R303.1.1.1, R402.2.2,
                               Table R405.5.2(1)

S

SCOPE R101.2
SERVICE HOT WATER
   Requirements R403.5
SHEATHING, INSULATING
   (see INSULATING SHEATHING)
SHGC
   (see SOLAR HEAT GAIN COEFFICIENT)
SHUTOFF DAMPERS R403.6
**SIMULATED PERFORMANCE
ALTERNATIVE** R405
   Documentation R405.4
   Mandatory requirements R405.2
   Performance-based compliance R405.3
   Report R405.4.2
   Software tools R405.4.1

SIZING
   Equipment and system R405.6.1
SKYLIGHTS R303.1.3, R402.1.2, R402.3,
                                Table R405.5.2(1)
   Defined R202
SNOW MELT SYSTEM CONTROLS R403.9
**SOLAR HEAT GAIN COEFFICIENT
(SHGC)** R102.1.3, R104.2,
                      Table R303.1.3(3), R401.3,
              Table R402.1.2, R402.1.4, R402.3.2,
                  R402.3.3, R402.3.6, R402.5
   Defined R202
STANDARD REFERENCE DESIGN
   Defined R202
   Requirements . . . . R405, Table R405.5.2(1)
STANDARDS, REFERENCED . . . . . R106, Chapter 5
STEEL FRAMING R402.2.6
STOP WORK ORDER R108
   Authority R108.1
   Emergencies R108.3
   Failure to comply R108.4
   Issuance R108.2
SUNROOM . . R402.2.13, R402.3.5, Table R405.5.2(1)
   Defined R202
   Insulation R402.2.13
SWIMMING POOLS R403.9

T

THERMAL CONDUCTANCE (see C-Factor)
THERMAL ISOLATION R402.2.12, R402.3.5,
                                  Table R405.5.2(1)
   Defined R202
THERMAL MASS (see MASS)
THERMAL RESISTANCE (see R-VALUE)
THERMAL TRANSMITTANCE (see U-FACTOR)
THERMOSTAT
   Defined R202
   Controls R403.1
   Programmable R403.1.1
TIME SWITCHES R403.9.2
TOTAL BUILDING PERFORMANCE
   Residential R405
TOWNHOUSE (see RESIDENTIAL BUILDINGS)
TROPICAL CLIMATE ZONES R301.4, R401.2.1

U

U-FACTOR
   Alternative . . . . R402.1.3, Table R402.1.3, R402.1.4
   Default door . . . . Table R303.1.3(2)

INDEX

Default glazed fenestration...... Table R303.1.3(1)
Defined R202, R402.3.1, R402.5
Skylights . Table R402.1.1, Table R402.1.3, R402.3.5
Sunroom............................R402.3.5

V

VALIDITY................................R105
VAPOR RETARDER.....................R402.1.1
VENTILATION.................. R401.2.1, R403.6,
           Table R403.6.1, Table R405.5.2(1)
Defined R202
VISIBLE TRANSMITTANCE (VT)
Default glazed fenestration...... Table R303.1.3(3)
Defined R202

W

WALL
Above-grade, defined.....................R202
   Standard reference design .... Table R405.5.2(1)
Basement, defined R202
   Installation........................R402.2.9
   Standard reference design .... Table R405.5.2(1)
Crawlspace, defined.....................R202
   Installation.......................R402.2.11
   Standard reference design .... Table R405.5.2(1)
Exterior, defined R202
Mass.................................R402.2.5
Steel-frame............ R402.2.6, Table R402.2.6
With partial structural sheathing R402.2.7
**WALLS (see EXTERIOR WALLS AND
BUILDING THERMAL ENVELOPE)**
**WALLS ADJACENT TO UNCONDITIONED
SPACE (see BUILDING THERMAL ENVELOPE)**
WATER HEATING......... R401.3, R403.4, R405.1,
                            Table R405.5.2(1)
**WHOLE HOUSE MECHANICAL
VENTILATION SYSTEM**
Defined................................R202
System fan efficacy R403.6.1
**WINDOW AREA
(see FENESTRATION and GLAZING AREA)**

EDITORIAL CHANGES – SECOND PRINTING

Page R-29, Section R401.2.1: Item 9 now reads . . .9. Bedrooms with exterior walls facing two different directions have operable fenestration.

Page R-32, Section R402.3.2: Exception, line 4 now reads . . .Table R402.1.2.

Page R-37, Section R403.10: line 4 now reads . . .through R403.10.3.

Page R-37, Section R403.10.1 has been deleted.

Page R-37, Section R403.10.2 title now reads . . .R403.10.1 Heaters

Page R-37, Section R403.10.3 title now reads . . .R403.10.2 Time switches

Page R-37, Section R403.10.4 title now reads . . .R403.10.3 Covers

Page R-39, TABLE R405.5.2(1): column 2, row 6 now reads . . .Emittance = 0.90

Page R-41, Section R406.2 now reads . . .R406.2 Mandatory requirements. Compliance with this section requires that the provisions identified in Sections R401 through R404 labeled as "mandatory" and Section R403.5.3 be met. The building thermal envelope shall be greater than or equal to levels of efficiency and Solar Heat Gain Coefficient in Table 402.1.1 or 402.1.3 of the 2009 *International Energy Conservation Code*.

Page R-44, Section R503.1.1.1: line 6 now reads . . .in Table R402.1.2.

Page R-50, TABLE RA301.1(3): column 1 now reads . . .CARBON MONOXIDE LEVEL (ppm)

EDITORIAL CHANGES – THIRD PRINTING

Page R-8, Section R202: HISTORIC BUILDING now reads . . .

HISTORIC BUILDING. Any building or structure that is one or more of the following:

1. Listed, or certified as eligible for listing by the State Historic Preservation Officer or the Keeper of the National Register of Historic Places, in the National Register of Historic Places.

2. Designated as historic under an applicable state or local law.

3. Certified as a contributing resource within a National Register-listed, state-designated or locally designated historic district.

Page R-26, Section R303.1.3 now reads as shown:

R303.1.3 Fenestration product rating. U-factors of fenestration products (windows, doors and skylights) shall be determined in accordance with NFRC 100.

Exception: Where required, garage door U-factors shall be determined in accordance with either NFRC 100 or ANSI/DASMA 105.

U-factors shall be determined by an accredited, independent laboratory, and *labeled* and certified by the manufacturer.

Products lacking such a *labeled* U-factor shall be assigned a default U-factor from Table R303.1.3(1) or R303.1.3(2). The solar heat gain coefficient (SHGC) and *visible transmittance* (VT) of glazed fenestration products (windows, glazed doors and skylights) shall be determined in accordance with NFRC 200 by an accredited, independent laboratory, and *labeled* and certified by the manufacturer. Products lacking such a *labeled* SHGC or VT shall be assigned a default SHGC or VT from Table R303.1.3(3).

Page R-29, Section R401.2.1: Item 9, lines 3 and 4 now read . . . directions have operable fenestration on exterior walls facing two directions.

Page R-32, Section R402.3: line 3 now reads . . .Sections R402.3.1 through R402.3.5.

Page R-33, Section R402.4: line 4 now reads . . .R402.4.5.

Page R-35, Section R403.3.3 Exception now reads as shown:

2. Postconstruction test: Total leakage shall be measured with a pressure differential of 0.1 inch w.g. (25 Pa) across the entire system, including the manufacturer's air handler enclosure. Registers shall be taped or otherwise sealed during the test.

Exception: A duct air leakage test shall not be required where the ducts and air handlers are located entirely within the building thermal envelope.

Page R-35, Section R403.5: line 3 now reads . . .with Sections R403.5.1 through R403.5.4.

Page R-39, Table R405.5.2: column 2, row 26, line 1 now reads . . . Total area[h] =

Page R-41, Table R405.5.2: Note h has been added.

Page R-42, Section R406.7.1, Item 1 now reads as shown:

1. Computer generation of the *ERI reference design* using only the input for the *rated design*.

    The calculation procedure shall not allow the user to directly modify the building component characteristics of the *ERI reference design*.

Page R-43, Section R502.1.1.1: Exception, line 4 now reads . . .in Section 402.1.5, of the existing building and the

Page R-44, Section R503.1.1: line 4 now reads . . .R402.2.13, R402.3.1, R402.3.2, R402.4.3 and R402.4.4.

At ICC Plan Review Services, our code experts love a good challenge

Like reviewing your most complex plan reviews.

Our licensed and ICC-certified code experts specialize in delivering the most accurate code analysis to your most complex and complicated projects. After all, no other plan review firm has as much expertise in **ALL** the International Codes (I-Codes) as the code experts at ICC Plan Review Services.

For I-Codes expertise you can trust, visit the ICC Plan Review web page at **www.iccsafe.org/PLRF2** or call David Hunter at 888-422-7233, ext. 5577.

14-09332

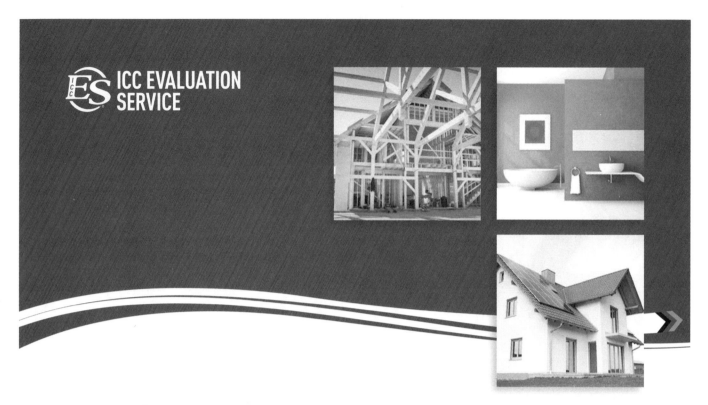

APPROVE WITH
CONFIDENCE

- ICC-ES® Evaluation Reports are the most widely accepted and trusted in the nation.
- ICC-ES is dedicated to providing evaluation reports, PMG and Building Product Listings to the highest quality and technical excellence.
- ICC-ES is a subsidiary of ICC®, the publisher of the IBC®, IRC®, IPC® and other I-Codes®.

We do thorough evaluations. You approve with confidence.

Look for the ICC-ES marks of conformity before approving for installation

 Code Officials — Contact Us With Your Questions About ICC-ES Reports at:
800-423-6587 ext. ES help (374357) or email EShelp@icc-es.org
www.icc-es.org

Subsidiary of

People Helping People Build a Safer World®

The International Code Council's Training and Education Department

Industry professionals look to ICC to provide the critical knowledge and experiences necessary to excel in today's challenging world. ICC's educational programs cover a broad spectrum of code and construction related topics, offering some of the highest quality instruction direct from the source.

INSTITUTES
Acquire Skills, Share Experiences, Earn CEUs
ICC's training institutes offer a comprehensive education experience and a great way to earn much needed CEUs. Learn best practices from the leading experts in the field, build your network of professional contacts and share experiences. Institutes are held across the country throughout the year. To see the full schedule go to www.iccsafe.org/training.

HIRE ICC TO TEACH
Bring ICC On-Site and Earn CEUs
Give your group the confidence they need to meet tough challenges head-on so they can reach their full potential. ICC's course catalog contains a wide variety of educational topics that are available for contract at a location of your choice. Customized training is also available to fit your needs. For our full course catalogue go to www.iccsafe.org/hireicc.

ICC ONLINE CAMPUS
Earn CEUs at Your Own Pace
Online courses offer access to a comprehensive training and education portal, providing you with an effective and convenient tool to enhance your professional skills. Courses are available anytime for 99 days following registration. For a quick and easy way to earn CEUs towards your certification, try our Online Certifications Renewal section. Go to www.iccsafe.org/onlinecampus for our full offerings.

For more information about ICC's Training please contact us at
888-ICC-SAFE (422-7233) ext. 33818 or email us at **icctraining@iccsafe.org**.

ICC INTERNATIONAL CODE COUNCIL

People Helping People Build a Safer World®

Growing your career is what ICC Membership is all about

As your building career grows, so does the need to expand your code knowledge and job skills. Whether you're seeking a higher level of certification or professional quality training, Membership in ICC offers the best in I-Code resources and training for growing your building career today and for the future.

- **Learn** new job skills to prepare for a higher level of responsibility within your organization
- **Improve** your code knowledge to keep pace with the latest International Codes (I-Codes)
- **Achieve** additional ICC Certifications to open the door to better job opportunities

Plus, an affordable ICC Membership provides exclusive Member-only benefits including:

- Free code opinions from I-Code experts
- Free I-Code book(s) to new Members*
- Access to employment opportunities in the ICC Career Center
- Discounts on professional training & Certification renewal exams
- Savings of up to 25% off on code books & training materials
- Free benefits - Governmental Members: Your staff can receive free ICC benefits too*
- And much more!

Join the International Code Council (ICC) and start growing your building career now!
Visit our Member page at **www.iccsafe.org/membership** for an application.

*Some restrictions apply. Speak with an ICC Member Services Representative for details.

14-09333